SPACE SCIENCE, EXPLORATION AND POLICIES SERIES

FUTURE U.S. SPACE LAUNCH CAPABILITIES

Space Science, Exploration and Policies Series

Progress in Dark Matter Research
J. Val Blain (Editor)
2005. ISBN 1-59454-248-1

Space Science: New Research
Nick S. Maravell (Editor)
2006. ISBN 1-60021-005-8

Space Policy and Exploration
William N. Callmers (Editor)
2008. ISBN 978-1-60456-448-8

Space Commercialization and the Development of Space Law from a Chinese Legal Perspective
Yun Zhao
2009. ISBN 978-1-60692-244-6

Next Generation of Human Space Flight Systems
Alfred T. Chesley (Editor)
2009. ISBN 978-1-60692-726-7

Smaller Satellites Operations Near Geostationary Orbit
Matthew T. Erdner
2009. ISBN 978-1-60741-181-9

Environmental Satellites: Weather and Environmental Information Systems
Vincent L. Webber (Editor)
2009. ISBN 978-1-60692-984-1

Nutritional Biochemistry of Space Flight
Scott M. Smith, Sara R. Zwart, Vickie Kloeris and Martina Heer
2009. ISBN 978-1-60741-641-8

Space Exploration Research
John H. Denis and Paul D. Aldridge (Editors)
2009. ISBN: 978-1-60692-264-4

Black Holes and Galaxy Formation
Adonis D. Wachter and Raphael J. Propst (Editors)
2010. ISBN: 978-1-60741-703-3

The International Space Station
David E. Jamison (Editor)
2010. ISBN: 978-1-60692-322-1

Future U.S. Space Launch Capabilities
Franz Lojdahl (Editor)
2010. ISBN: 978-1-60741-384-4

SPACE SCIENCE, EXPLORATION AND POLICIES SERIES

FUTURE U.S. SPACE LAUNCH CAPABILITIES

FRANZ LOJDAHL
EDITOR

Nova Science Publishers, Inc.
New York

LIBRARY OF CONGRESS CATALOGING-IN-PUBLICATION DATA

Future U.S. space launch capabilities / editor, Franz Lojdahl.
p. cm.
Includes index.
ISBN 978-1-60741-384-4 (hardcover)
1. Space race--United States. 2. Astronautics--United States. 3. Outer space--Exploration--United States. 4. United States. National Aeronautics and Space Administration--Finance. I. Lojdahl, Franz.
TL789.8.U5F85 2009
629.40973--dc22 2009035604

Published by Nova Science Publishers, Inc. ✝ New York

CONTENTS

PREFACE

This new book presents a wide spectrum of in-depth analysis detailing the U.S. space program including policy, the space stations, the shuttles and space exploration. The "space age" began on October 4th 1957, when the Soviet Union launched Sputnik, the world's first artificial satellite. Some U.S. policymakers, concerned about the USSR's ability to launch a satellite, thought Sputnik might be an indication that the U.S. was trailing behind the USSR in science and technology. The Cold War also led some to perceive the Sputnik launch as a possible precursor to nuclear attack. In response to the "Sputnik moment", the U.S. government undertook several policy actions, including the establishment of the National Aeronautics and Space Administration (NASA), the enhancement of research funding, and reformation of science, technology, engineering and mathematics (STEM) education policy. Also following the "Sputnik moment," a set of fundamental factors gave "importance, urgency, and inevitability to the advancement of space technology." These four factors, discussed in this book, include the compelling need to explore and discover, national defense, prestige and confidence in the U.S. scientific, technological, industrial and military systems and scientific observation and experimentation to add to our knowledge and understanding of the Earth, solar system, and universe.

Chapter 1 - A Presidential directive issued on January 14, 2004—called the new Vision for Space Exploration (VSE)—set out goals for future exploration of the solar system using manned spacecraft. Those goals included returning to the moon no later than 2020. Although sufficient capabilities exist to meet the projected needs of both the U.S. commercial sector and the government for launching unmanned payloads into space through 2020, that is not the case for manned space flight. The proposed return to the moon called for under the VSE and now planned by the National Aeronautics and Space Administration (NASA) could require the development of the capacity to launch payloads weighing more than 100 metric tons (mt).[1] No launch vehicles currently exist that can handle payloads weighing more than about 25 mt. Thus, NASA's plans for manned space flight beyond low earth orbit (LEO) could require a significant increase in launch capability.[2] How that capability could be provided and at what cost are the focal points of this study.

In considering manned lunar missions, the Congressional Budget Office (CBO) explored alternatives that would use existing launch vehicles; those that would require minor modifications to the designs of existing launchers (termed "close derivatives"); as well as those that would call for major modifications to existing vehicle designs to develop essentially new and much more capable launchers.[3] All of the alternatives would require multiple launches to assemble in LEO the fuel and hardware needed to fly to the moon. Under the alternatives, CBO estimates, the costs to develop and procure launch vehicles that could support a manned lunar mission in 2018 (under a more ambitious schedule) would range from $26 billion to $38 billion. NASA's projection of funding is $30 billion. The use of less capable existing or close-derivative launchers could be less costly but would require up to eight launches to assemble a single lunar mission. Using the new and more capable launchers considered by CBO would be more costly but could reduce to two the number of launches needed per lunar mission. The greater the number of launches needed to assemble a mission, the greater the complexity of the mission— including both the need to perform on-orbit assembly of the mission's components and the risk that at least one launch would fail, putting the success of the mission at risk. Thus, there is a trade-off between the overall costs of launch vehicles and the risk of mission failure.

Chapter 2 - The Department of Defense (DOD) plans to spend over $27 billion acquiring launch services through the Evolved Expendable Launch Vehicle (EELV) program over the next 12 years. The EELV program uses two families of commercially owned and operated vehicles to launch satellites.

Partly because the commercial space market did not develop as expected, the EELV program has undergone significant changes. These include: adoption of a new acquisition strategy in 2005 that sought to ensure the viability of the two EELV launch vehicle providers, Boeing and Lockheed Martin; the subsequent decision by those two companies to form a joint venture called the United Launch Alliance (ULA); and a 10- year increase in the life of the program. In light of these changes, GAO was asked to (1) determine what uncertainties DOD faces in the EELV program and in the transition to ULA, and (2) assess how DOD is positioned to manage and oversee the effort. To accomplish this, GAO reviewed a wide variety of DOD documents and interviewed DOD and program officials.

Chapter 3 - Launching satellites into orbit, once the exclusive domain of the U.S. and Soviet governments, today is an industry in which companies in the United States, Europe, China, Russia, Ukraine, Japan, and India compete. In the United States, the National Aeronautics and Space Administration (NASA) owns and launches its space shuttle. Private sector companies provide launch services for other NASA launches and most Department of Defense (DOD) launches. Commercial customers purchase launch services from the U.S. companies or their competitors. Since the early 1980s, Congress and successive Administrations have taken actions, including passing several laws, to facilitate the U.S. commercial space launch services business. The Federal Aviation Administration (FAA) regulates the industry.

Forecasts in the 1990s suggesting significant increases in launch demand sparked plans to develop new launch v ehicles. NASA and DOD created government-industry partnerships to develop new reusable launch vehicles (RLVs) and "evolved" expendable launch vehicles (EELVs), respectively. (The space shuttle is the only RLV today. All other launch vehicles are "expendable" — they can only be used once). Several U.S. private sector companies began developing their own launch vehicles. Projections for launch services demand declined dramatically beginning in 1999, however. NASA's efforts to develop a new RLV to replace the shuttle faltered. DOD's new EELVs (Atlas V and Delta IV) began service, but, with reduced demand, the companies that build them (Lockheed Martin and Boeing) want more DOD funding to defray their costs. In 2005, the two companies announced plans to merge their EELV launch services for U.S. government customers. The joint venture, if approved by regulatory authorities, would be named the United Launch Alliance. Commercial launch services would not be affected.

Congress is debating the future of the space shuttle, which returned to flight in July 2005 after a two and one-half year hiatus following the 2003 *Columbia* tragedy. President Bush directed NASA to terminate the shuttle in 2010, but some want it to continue until a replacement is available. NASA also is embarking on a program to build two new "shuttle-derived" launch vehicles to implement the President's "Vision for Space Exploration" to return astronauts to the Moon and someday send them to Mars. The FY2007 request includes $837 million for the first of these new vehicles, the Crew Launch Vehicle.

In October 2004, Burt Rutan's SpaceShipOne suborbital spacecraft won the $10 million Ansari X-prize. Some believe this heralds an era of comparatively affordable space tourism. Congress passed a law in 2004 (P.L. 108-492) that creates a regulatory environment for space tourism. Concerns that China benefitted militarily from knowledge gained through commercial satellite launches in the 1990s led to changes in U.S. satellite export policy. The changes, especially returning control over such exports to the State Department from the Commerce Department, remain controversial because of what some claim is a negative impact on U.S. satellite manufacturing companies whose clients may choose European suppliers to avoid the U.S. export control regulations.

Chapter 4 - Congress continues to debate NASA's International Space Station (ISS), a permanently occupied facility in Earth orbit where astronauts live and conduct research. Congress appropriated approximately $35 billion for the program from FY1985-2005. The initial FY2006 ISS request was $2.180 billion: $1.857 billion for construction and operations and $324 million for research to be conducted by ISS crews. In a July budget amendment, NASA transferred $168 million for ISS Crew/Cargo Services to another part of the NASA budget and reduced the ISS request commensurately. The final version of the FY2006 appropriations bill that includes NASA (H.R. 2862) cuts $80 million from the originally submitted budget, and NASA now plans to spend $306 million, instead of $324 million, on ISS research in FY2006.

The space station is being assembled in Earth orbit. ISS segments, crews, and cargo are taken into orbit by Russian or U.S. spacecraft. ISS has been permanently occupied by successive "Expedition" crews rotating on 4-6 month shifts since November 2000. "Expedition 12" is now aboard. Cost growth and schedule delays have characterized the program since its inception. The grounding of the space shuttle fleet after the 2003 *Columbia* tragedy and the July 2005 *Discovery* "Return to Flight" mission is further affecting schedule, operations, and cost. Most of the remaining ISS segments are designed to be launched by the shuttle and construction therefore is suspended. President Bush's January 2004 "Vision for Space Exploration" also is affecting the ISS program. He directed that the shuttle program be terminated in 2010, and changed the focus of ISS-based research to only that which supports his "Moon/Mars" Vision instead of the broadlybased program that was planned.

Canada, Japan, and several European countries became partners with NASA in building the space station in 1988; Russia joined in 1993. Except for money paid to Russia, there is no exchange of funds among the partners. Europe, Canada, and Japan collectively expect to spend about $11 billion of their own money. A reliable figure for Russian expenditures is not available.

In 1993, when the current space station design was adopted, NASA said it would cost $17.4 billion for construction (not including launch or other costs). That estimate grew to $24.1-$26.4 billion, leading Congress to legislate spending caps on part of the program in 2000. The estimate then grew by almost another $5 billion, leading NASA (at White House direction) to cancel or indefinitely defer some hardware to stay within the cap. NASA exceeded the cap in FY2005, however.

Controversial since the program began in 1984, the space station has been repeatedly redesigned and rescheduled, often for costgrowth reasons. Congress has been concerned about the space station for that and other reasons. Twenty-two attempts to terminate the program in NASA funding bills were defeated, however (3 in the 106th Congress, 4 in the 105th Congress, 5 in the 104th, 5 in the 103rd, and 5 in the 102nd). Three other attempts in broader legislation in the 103rd Congress also failed.

Current congressional debate focuses on the impact of space shuttle-related delays, and the future of ISS in light of President Bush's new exploration initiative.

In: Future U.S. Space Launch Capabilities
Editors: Franz Lojdahl ISBN: 978-1-60741-384-4

Chapter 1

ALTERNATIVES FOR FUTURE U.S. SPACE-LAUNCH CAPABILITIES

Congressional Budget Office

INTRODUCTION AND SUMMARY

A Presidential directive issued on January 14, 2004—called the new Vision for Space Exploration (VSE)—set out goals for future exploration of the solar system using manned spacecraft. Those goals included returning to the moon no later than 2020. Although sufficient capabilities exist to meet the projected needs of both the U.S. commercial sector and the government for launching unmanned payloads into space through 2020, that is not the case for manned space flight. The proposed return to the moon called for under the VSE and now planned by the National Aeronautics and Space Administration (NASA) could require the development of the capacity to launch payloads weighing more than 100 metric tons (mt).[1] No launch vehicles currently exist that can handle payloads weighing more than about 25 mt. Thus, NASA's plans for manned space flight beyond low earth orbit (LEO) could require a significant increase in launch capability.[2] How that capability could be provided and at what cost are the focal points of this study.

In considering manned lunar missions, the Congressional Budget Office (CBO) explored alternatives that would use existing launch vehicles; those that would require minor modifications to the designs of existing launchers (termed "close derivatives"); as well as those that would call for major modifications to existing vehicle designs to develop essentially new and much more capable launchers.[3] All of the alternatives would require multiple launches to assemble in LEO the fuel and hardware needed to fly to the moon. Under the alternatives, CBO estimates, the costs to develop and procure launch vehicles that could support a manned lunar mission in 2018 (under a more ambitious schedule) would range from $26 billion to $38 billion. NASA's projection of funding is $30 billion. The use of less capable existing or close-derivative launchers could be less costly but would require up to eight launches to assemble a single lunar mission. Using the new and more capable launchers considered by CBO would be more costly but could reduce to two the number of launches needed per lunar mission. The greater the number of launches needed to assemble a mission, the greater the complexity of the mission—including both the need to perform on-orbit assembly of the mission's components and the risk that at least one launch would fail, putting the success of the mission at risk. Thus, there is a trade-off between the overall costs of launch vehicles and the risk of mission failure.

Current Launch Capabilities and Projected Worldwide Demand

Excluding manned flight beyond LEO, U.S. needs for launch vehicles involve putting into orbit payloads (both commercial and governmental in origin) that weigh less than 25 metric tons, with the majority weighing less than 12 mt. Typical payloads consist of satellites (usually designed to track weather, facilitate communications, aid scientific research, or conduct surveillance) or mission support (such as consumables, parts, or manpower) for the International Space Station, or ISS. Current estimates from various sources of worldwide launch capacity for payloads weighing less than about 25 mt range from approximately 120 to 150 launches per year. According to projections obtained by CBO, worldwide demand for launching such payloads through 2020 ranges between 70 and 80 launches a year (see Summary Figure 1). Projected capacity is an estimate of the maximum number of launches that existing launch pads and launch vehicle manufacturing facilities can support if fully manned and funded. Projected demand includes all launches now expected worldwide for either commercial or governmental purposes. The projections obtained by CBO indicate that maximum worldwide launch capacity for payloads of less than 25 mt will exceed demand by about 50 percent to as much as 100 percent.

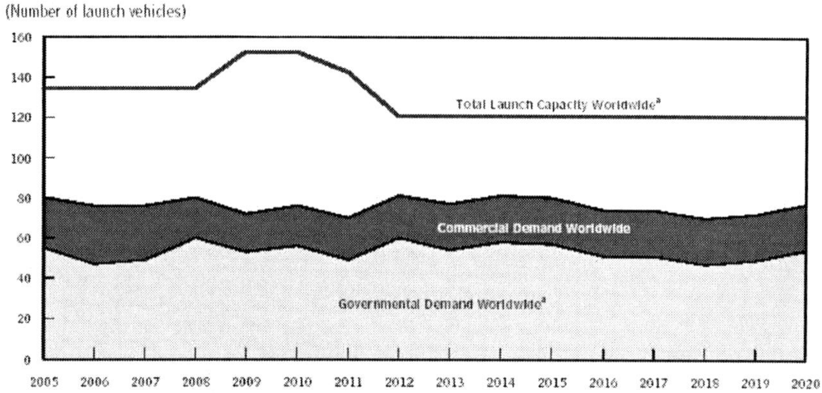

Source: Congressional Budget Office based on data provided by the Federal Aviation Administration and the Futron Corporation.

Note: Capacity is the number of launches that existing infrastructure and production facilities can support if fully manned and funded. Demand is either the number of launches required on historical launch manifests or current projections of future launch manifests. For a variety of reasons, therefore (because of delays in the availability of payloads, for instance), the number of actual launches usually ends up being less than the demand reflected on manifests.

a. Excludes the National Aeronautics and Space Administration's Vision for Space Exploration initiative.

Summary Figure 1. Worldwide Capacity and Demand for Launch Services

Launch Needs for Manned Lunar Missions

The flights to the moon conducted during the Apollo program (from 1966 to 1972) required the capacity to launch about 140 mt into LEO. That capability was provided by a single, expendable launch vehicle, the Saturn V. The details of the lunar missions being planned by NASA under the VSE could be more ambitious than those conducted under the Apollo program. (One motivation for resuming lunar missions is to help enable eventual manned space flight beyond the moon—including to Mars.) NASA also plans to use the spacecraft that would be used to fly to and return from the moon—the crew exploration vehicle, or CEV—to service the ISS after the retirement of the space shuttle in 2010. NASA has estimated that the net effect of those changes relative to Saturn/Apollo might increase the weight of a manned lunar mission by about 10 percent.

The alternatives considered by CBO for returning humans to the moon fall into two categories: those that would use existing heavy launchers and those that would require developing new launch systems. The new launch systems considered by CBO include close derivatives of existing systems—launchers that could lift somewhat heavier payloads than current systems—as well as new, larger launchers, referred to as super heavies, that could lift payloads weighing 100 mt or more. Generally, the alternatives considered by CBO have also been considered by NASA or its contractors, including Boeing, Lockheed Martin, and ATK Thiokol, as NASA continues to develop its plans for returning humans to the moon.

Existing Heavy-Lift Launch Vehicles

The existing U.S. heavy-lift launch systems considered by CBO include NASA's space shuttle, Boeing's Delta IV Heavy (D4H) variant of the U.S. Air Force's set of evolved expendable launch vehicles (EELVs), and the most capable variant of Lockheed Martin's Atlas V Medium (A5M) EELV.[4] The Delta IV Heavy has a lift capacity of 25 mt, the most capable variant of the Atlas V Medium has a lift capacity of 20.5 mt, and the space shuttle can lift a payload weighing 18 mt.[5] Therefore, using existing U.S. heavy-lift launch systems would require six to eight launches to propel into LEO all of the mass necessary to execute a manned lunar mission.

Although in the Apollo program the payload for lunar exploration was lifted into LEO with one launch, NASA considered options for the VSE program that require multiple launches. The use of multiple launches has wide-ranging implications for both safety and cost savings. In particular, such a strategy could theoretically permit the use of existing launchers rather than require the development of a new launch vehicle with lift capability exceeding 100 mt. (That would translate into less capable but individually cheaper launch vehicles.) Using multiple launches would also mean, again theoretically, that a single launch failure would not cause the failure of an entire mission. Further, multiple launches could permit the crew to be lifted on a launcher designed to be highly reliable, while less reliable launchers were used to lift the other, heavier elements of the payload, or cargo.

There are, however, potential drawbacks associated with using multiple launches. Multiple launches could require time to execute (perhaps several months), during which subcomponents of the full payload, some likely containing cryogenic fuels, must remain in LEO. Notwithstanding improved techniques for insulating cryogenic fuels, on-orbit storage of such fuels for many months would be a challenge (because of leakage) and would require that fuel in excess of the amounts actually burned be lifted into LEO, thereby potentially increasing the total number of launches needed. The subcomponents would also have to be assembled into a complete spacecraft in orbit, potentially requiring the development of a capability for reliable and fully automated rendezvous, docking, and on-orbit assembly that the United States has never attempted.

Relying on multiple launches could also complicate mission planning because the more launches that needed to be executed to complete a mission, the less the probability that all of those launches would succeed. A launch failure or failure of an on-orbit rendezvous and docking would render a payload subcomponent unusable, and which subcomponent might fail could not be predicted a priori. Thus, a complete set of additional payload sub-components might be needed to ensure completion of a successful mission.

New Launch Vehicles: Close Derivatives of Existing Systems

Although representative close derivatives selected by CBO have the capability to lift payloads two to four times heavier than existing launchers, that increase in payload weight would be achieved by making major modifications to just a few components of an existing launch system rather than major modifications to nearly all of its components. In this study, CBO analyzed the cost and performance of close-derivative launch vehicles that could be used to execute a manned lunar mission similar to those considered by NASA and members of the U.S. space-launch industry. The close derivatives considered by CBO included modified versions of the space shuttle and its components, as well as modified versions of both the Atlas and Delta versions of the EELV. In all cases, the launch vehicles used to lift a spacecraft and its crew into LEO, called crew carriers, are distinct from cargo carriers, which are used to lift the remaining payload.

The specific close-derivative launch vehicles considered by CBO include the following:

- **The Five-Segment "Stick" (crew carrier/lift capacity, 26 mt).** This launch vehicle's antecedent is based on the four-segment solid rocket booster (SRB), a pair of which now help lift the shuttle into orbit. (The main propulsion system for the shuttle is composed of three "sticks," two four-segment SRBs attached to the central external tank, or ET—the third stick.) The SRBs are composed of individual cylindrical sections, each of which contains a casting of solid fuel. The single five-segment stick considered here would consist of five such sections, which constitute the first stage of the rocket, as well as a new second stage (with an independent liquid-fuel propulsion system) mounted on top of the SRB. That additional stage is needed to provide the capability to lift 26 mt into LEO.

- **The Side-Mount (cargo carrier/lift capacity, 77 mt).** Also conceptually similar to the space shuttle, the Side-Mount would use the space shuttle propulsion system with a cargo canister replacing the airplane-like orbiter. The payload that the side-mount could lift into LEO would be about 77 mt.

- **The Delta IV Heavy with Modifications, or D4H with Mods (crew carrier/lift capacity, 21 mt).** Based on the existing Delta IV Heavy developed by Boeing, the D4H with mods would be able to lift about 21 mt into LEO.

- **The Delta IV Medium Plus, or D4M+ (crew carrier/lift capacity, 18 mt).** Relative to the single-stick Delta IV Medium (D4M) EELV, from which it is derived, the D4M+ would have an expanded first-stage central core with additional engines and an expanded second stage. Those modifications would enable the D4M+ to lift about 18 mt into LEO.

- **The Delta IV Heavy Plus, or D4H+ (cargo carrier/lift capacity, 40 mt).** The three-stick D4H is built using three of the first-stage central cores used for the D4M. Similarly, the D4H+ would use three of the central cores of the D4M+, enabling it to lift about 40 mt of cargo into LEO.

- **The Atlas V Medium Plus, or A5M+ (crew carrier/lift capacity, 24 mt).** This single-stick vehicle is based on the Atlas V Medium (A5M) EELV, developed by Lockheed Martin. Relative to the existing A5M launcher, the A5M+ would have a larger first-stage central core and an upgraded second stage providing the capability to lift about 24 mt into LEO.

- **The Atlas V Heavy Plus, or A5H+ (cargo carrier/lift capacity, 74 mt).** This Atlas-derived cargo carrier would be based on the A5M+ crew carrier. The first stage of the A5M+ would serve as the basis for this three-stick vehicle. Relative to the A5M+, the A5H+ would also feature additional engines in its second stage. It would have the capability to lift about 74 mt into LEO.

New Launch Vehicles: Super Heavies

The super-heavy launch vehicles considered by CBO could lift payloads of 100 metric tons or more into low earth orbit. In general, those prospective launchers would require significant and relatively expensive upgrades to many—if not all—of the major systems that compose existing launchers. CBO considered super-heavy launch vehicles that would use components of the space shuttle, as well as launchers that would represent new and much more capable designs incorporating selected components—primarily engines—used on the current EELVs.

The specific super-heavy launchers considered by CBO include the following:

- **The Top-Mounted, Shuttle-Derived Super Heavy (cargo carrier/lift capacity, 125 mt).** In the January 2006 release of the President's 2007 budget, this cargo launch vehicle was identified as NASA's recommended solution for supporting manned lunar missions.[6] Because many of its major components would be drawn from the space shuttle, CBO refers to the vehicle as the shuttle-derived super heavy. It would be about the size of the Saturn V rocket used for the Apollo flights to the moon, and its payload to LEO would be about 125 mt. It would use two five-segment SRBs (as opposed to the shuttle's two four- segment SRBs). The five-segment SRBs would be mounted on the sides of an elongated version of the shuttle's existing external tank, which would be modified to house five expendable versions of the existing (reusable) space shuttle main engines (SSMEs).[7] A large cargo canister, or "shroud," would be mounted on top of the tank-SRB assembly.

- **The Longfellow (cargo carrier/lift capacity, 108 mt).** This launch vehicle—proposed by an industry consortium that includes Boeing, Lockheed Martin, and ATK Thiokol—would resemble the shuttle-derived super heavy, but several aspects of its design would be different. In particular, it would use less expensive engines in its first stage and would be taller to accommodate a second stage. The Longfellow would be able to lift about 108 mt into LEO.

- **The Delta Super Heavy (cargo carrier/lift capacity, 146 mt).** The closest antecedent of this vehicle—proposed by Boeing—is the D4H. This launcher, however, would incorporate many changes relative to either the existing D4H or the prospective D4H+. Those changes include the use of three expanded first- stage cores 8 meters in diameter. The largest of the super heavies considered by CBO, this launcher would be capable of lifting the largest payload into LEO—about 146 mt.

- **The Atlas Super Heavy (cargo carrier/lift capacity, 135 mt).** The closest antecedent of this vehicle—proposed by Lockheed Martin—would be the Atlas V Heavy Plus. Relative to that prospective launcher, the Atlas super heavy would have a substantially expanded 8.4-meter first-stage central core powered by five engines. The expanded central core would be surrounded by four of the smaller first-stage cores used in the A5H+'s triple-stick configuration. Thus, the Atlas super-heavy launcher's first stage would be a five-stick configuration with 13 engines. The launcher's second stage would be similar to that of the A5H+ but would incorporate a lengthened payload shroud. Those changes would result in a vehicle capable of lifting 135 mt into LEO.

Six Alternative Launch Programs

CBO has constructed projections of the budgetary resources needed through 2017 to develop and procure launch vehicles that would support the return to the moon called for under the VSE. The projections are built for a total of six alternative launch programs—three that would use close-derivative launchers and three that would use super-heavy launchers—as follows:

- **The "pure Atlas-derived" alternative.**[8] This program would use the A5M+ launcher as the crew carrier and the A5H+ launcher as the cargo carrier.

- **The "pure Delta-derived" alternative.** This program would feature the D4M+ as the crew carrier and the D4H+ as the cargo carrier.

- **The "pure shuttle-derived" alternative.** This program would use the five-segment stick as the crew carrier and the Side-Mount as the cargo carrier.

- **The "pure Atlas-antecedent" alternative.** This program would use the A5M+ launcher as the crew carrier and the Atlas super heavy launcher as the cargo carrier.

- **The "pure Delta-antecedent" alternative.** This program would use the D4M+ as the crew carrier and the Delta super heavy launcher as the cargo carrier.

- **The "shuttle-derived super-heavy" alternative.** This program would use the five-segment stick as the crew carrier and the top-mounted, shuttle-derived super- heavy launcher as the cargo carrier. With the release of the President's budget for fiscal year 2007, it was also NASA's choice for VSE launch vehicles.

Programmatic Assumptions and Cost Comparisons

In order to construct budgetary projections for the alternatives described above, CBO needed to make a variety of "programmatic" assumptions, such as when launch vehicles would first be used, whether they would be used to lift cargo or a spacecraft and its crew, and how many test flights would be necessary before a launcher was deemed ready for use. CBO used two sources for making those assumptions. First, to the extent that NASA's September 2005 Exploration Systems Architecture Study (ESAS) specified schedules and other program details, CBO used that study's assumptions—unless they were supplanted by materials the agency prepared to support the submission of the President's budget for 2007. In particular, justification materials supporting the 2007 budget indicate that NASA remains hopeful that it can meet the schedule outlined in the ESAS for conducting the first flight of the crew exploration vehicle, which is set for 2012, and the first manned lunar mission, which is planned for 2018. However, the agency recognizes that budget constraints might delay those events until 2014 and 2020, respectively. Therefore, CBO has constructed budgetary projections for the launch vehicle funding needed to execute both of those schedules. The programmatic assumptions that CBO made include the following:

- Certification of a launch vehicle—whether a crew carrier or a cargo carrier—for operational use would require three test or demonstration (test/demo) flights.
- Appropriations for hardware production and launch services would precede a launch by two years.

- The first operational flight of the CEV (which would be lifted into LEO by the crew carrier) would occur in either 2012 (under the more ambitious schedule) or 2014 (according to the less ambitious schedule). The first manned lunar mission would occur in either 2018 (the more ambitious schedule) or 2020 (the less ambitious schedule).
- Five ISS support flights per year would be conducted through 2016, beginning in either 2012 or 2014. Cumulative ISS support flights would therefore total 25 or 15. The launch vehicle for those flights would be the CEV launcher, regardless of whether or not the payload to the ISS included a crew.
- Two manned lunar missions would occur annually, beginning in either 2018 or 2020.
- No budgetary allowances for launch failure are made.

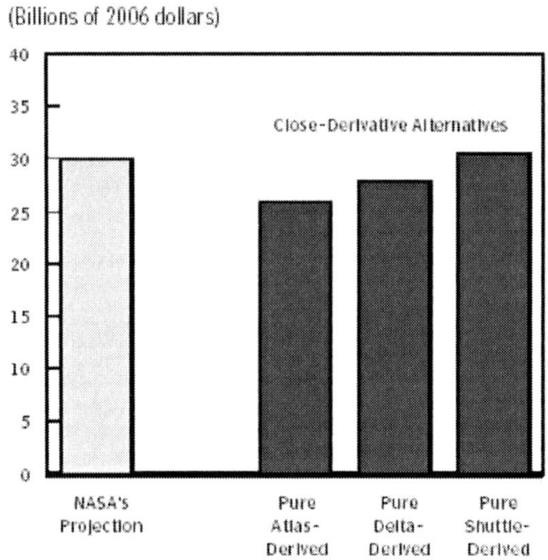

Source: Congressional Budget Office.

Notes: NASA = National Aeronautics and Space Administration.

The more ambitious schedule initiates International Space Station support in 2012 and the first lunar mission in 2018.

The term "pure" means that both the crew carrier and the cargo carrier are derived from vehicles in the same family of launchers.

The term "close derivative" describes new launchers that retain a close pedigree to existing systems.

Summary Figure 2. Costs Through 2017 of Using Close-Derivative Launchers to Conduct Manned Lunar Missions Under the More Ambitious Schedule

To execute the more ambitious schedule for returning humans to the moon, CBO estimates that total costs through 2017 for pursuing the three alternatives using close-derivative launch vehicles would range from about $26 billion for the pure Atlas-derived launch program to almost $31 billion for the pure shuttle-derived program (see Summary Figure 2). Executing the less ambitious schedule would cost $15 billion to $16 billion through 2017, CBO projects. In that case, the costs of executing the more ambitious schedule would simply be deferred to later years.

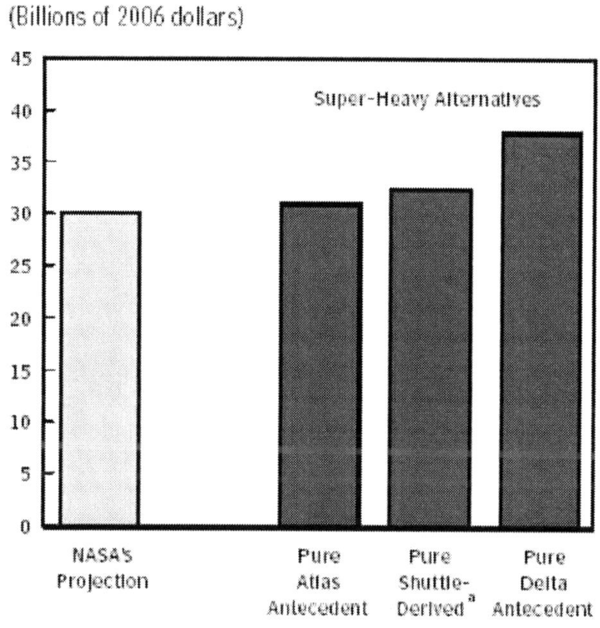

Source: Congressional Budget Office.

Notes: NASA = National Aeronautics and Space Administration.

The more ambitious schedule initiates International Space Station support in 2012 and the first lunar mission in 2018.

Each alternative includes the close-derivative crew carrier associated with the indicated family. The term "pure" means that both the crew carrier and the cargo carrier are derived from vehicles in the same family of launchers.

The term "close derivative" describes new launchers that retain a close pedigree to existing systems.

a. NASA's choice of crew launch vehicle and cargo carrier as outlined in the President's budget for 2007.

Summary Figure 3. Costs Through 2017 of Using Super-Heavy Launchers to Conduct Manned Lunar Missions Under the More Ambitious Schedule

To execute the more ambitious schedule using the three alternatives that incorporate super-heavy launchers, CBO estimates that costs through 2017 would range from about $31 billion to $38 billion (see Summary Figure 3). To execute the less ambitious schedule using super-heavy launchers, CBO estimates that costs through 2017 would range from about $20 billion to $26 billion. Again, executing the less ambitious schedule would simply defer costs to later years.

Both schedules imply a period when NASA could not support manned space flight with the proposed VSE vehicles after the retirement of the space shuttle in 2010. With the earlier, more ambitious schedule, the hiatus would be two years, and with the less ambitious schedule, it would be four years.

NASA's funding projections through 2017 to support the development and purchase of VSE launch vehicles total $30.1 billion. Through 2010, the projected amounts are about $1.0 billion annually. After 2010, annual amounts are about $3.5 billion. NASA's projection is constructed to support development and use of the shuttle-derived five-segment stick as the crew carrier and the similarly derived top-mounted super-heavy launcher as the cargo carrier—that is, the shuttle-derived super-heavy alternative considered by CBO. Through 2017, CBO's estimate of the budgetary resources needed to execute that alternative on the more ambitious schedule totals about $32 billion; CBO's estimate for executing the less ambitious schedule totals $20 billion.[9]

1. Current Capabilities and Plans

With the exception of plans for manned space flight beyond low earth orbit (LEO)—which have been called for under the Bush Administration's Vision for Space Exploration (VSE)—sufficient capabilities exist through 2020 to meet projected needs of both U.S. commercial and governmental launching of unmanned payloads into space.[10] Those capabilities involve launching payloads that weigh less than 25 metric tons (mt), with the majority weighing less than 12 mt. Manned space flight beyond LEO, however, such as the return to the moon proposed under the VSE and now planned by the National Aeronautics and Space Administration (NASA), could require the development of the capability to launch payloads that weigh more than 100 mt. No launch vehicles presently exist that can handle payloads weighing more than about 25 mt. Thus, using current launch vehicles for a manned lunar mission would require that in excess of 100 mt of hardware and fuel be divided into multiple payloads that could be launched separately and then assembled into a functional whole in LEO before the mission departed for the moon.

The Vision for Space Exploration, which was issued as a Presidential directive on January 14, 2004, outlined goals for future exploration of the solar system using manned spacecraft. Among those goals was a proposal to return humans to the moon no later than 2020. The ultimate goal of resuming lunar missions would be to help enable eventual manned space flight beyond the moon—including to Mars. During the Apollo program (in effect from 1966 to 1972), manned space flight to the moon required the capability to launch about 140 mt into LEO, which was provided by a single Saturn V launcher.

The details of the lunar missions being planned by NASA under the VSE could be more ambitious than those conducted under the Apollo program. In addition, NASA plans to use the crew exploration vehicle (CEV), the spacecraft that will be used to fly to and return from the moon and to service the International Space Station (ISS) after the retirement of the space shuttle in 2010. As will be discussed later, NASA has estimated that the net effect of those changes relative to Saturn/Apollo would increase the weight of a manned lunar mission by about 10 percent. Manned missions to Mars would, at a minimum, require launching payloads into LEO many times the weight required for lunar missions.

In this section of its analysis, the Congressional Budget Office (CBO) discusses projections that the Federal Aviation Administration (FAA) has made for commercial and governmental launches of payloads weighing less than 25 mt—that is, payloads not involving the VSE launch capabilities. Those projections indicate that currently existing launch capabilities should be sufficient to handle payloads weighing less than 25 mt through at least 2020. This chapter also discusses the issues that CBO considered in developing and assessing alternative launch systems that could potentially be used to execute the VSE. The specific alternatives that CBO developed are discussed in detail in Chapter 2.

Current Launch Capabilities and Projected Worldwide Demand Through 2020

Current estimates of worldwide launch capacity for pay loads that weigh less than about 25 metric tons range from about 120 to 150 launches per year. Current projections of worldwide demand for launching such payloads range between 70 and 80 launches a year (see Figure 1.1).[11] Projected capacity is an estimate of the maximum number of launches that existing launch pads and launch vehicle manufacturing facilities can support. Projected demand includes all launches now expected worldwide for either commercial or governmental purposes. Current projections indicate that maximum worldwide launch capacity for payloads of less than 25 mt will exceed demand by up to 100 percent for the foreseeable future.

Those projections of excess launch capacity may be underestimated, however. Data compiled by the FAA since 1993 indicate that actual launches of payloads— such as communications satellites—into geosynchronous orbit (GSO) range from about 60 percent to 86 percent of what has been forecast.[12] On average, actual launches conducted by the U.S. Air Force have been about 72 percent of its forecasts.[13] Projected launches exceed actual launches for a variety of reasons, including delays in the development and manufacture of the satellites scheduled for launch.

Worldwide demand for launches of payloads that weigh between 11.4 mt and 25 mt, referred to as heavy-launch demand by the FAA, is projected to be about 20 annually, much less than the estimated capacity (see Box 1.1 on page 4). Several current launch systems can accommodate the lifting of payloads at the upper end of that range (or beyond) into certain low earth orbits, including the space shuttle and the heavy-lift version of Boeing's Delta IV launcher.[14] Until its planned retirement in 2010, the space shuttle will be used almost exclusively to lift payloads associated with servicing and completing the construction of the ISS. Projected demand for Delta IV Heavy (D4H) launches is currently as high as two per year for U.S. governmental payloads. Launch vehicles with a maximum lift capability of between 11.4 mt and 20 mt include other versions of Boeing's Delta IV, versions of Lockheed Martin's Atlas V launcher, and a number of launchers manufactured abroad. The latter include the Ariane 5G used by the European Space Agency, the Chinese Long March 3B, and several variants of the Zenit and Proton launch systems now operated by the Single Economic Space.[15] The projected maximum capacity of those foreign systems, the majority of which can lift payloads no greater than about 18 mt, is somewhat greater than 40 launches annually.

The majority of the projected future worldwide demand for launches (about 72 percent) is associated with plans by governments, such as those of Russia, China, the United States, and members of the European Union, to place into orbit environmental-sensing satellites, other remote-sensing satellites (for example, synthetic aperture radars), and military-application satellites. With the exception of the period spanning 1997 to 2001—when launches of communications satellites from companies such as Iridium and Globalstar caused commercial demand to rise to about 42 percent of worldwide launches—this projection of continued dominance by governments for launch services is consistent with historical experience.

Although the number of satellites launched annually is projected to increase, the number of projected launches remains roughly constant.[16] That is because advances in electronics and composite materials are assumed to continue to reduce satellite mass, thereby allowing multiple satellites to be lifted into orbit during a single launch (known as multi-manifesting). That assumption also underlies the projection of the need for fewer launches of payloads that exceed 20 mt. (About 75 percent of payloads are lifted by launch vehicles that are not classified as heavy. At the upper end of the heavy range—that is, about 21 mt—projected demand through 2020 amounts to less than 3 percent of total demand.)

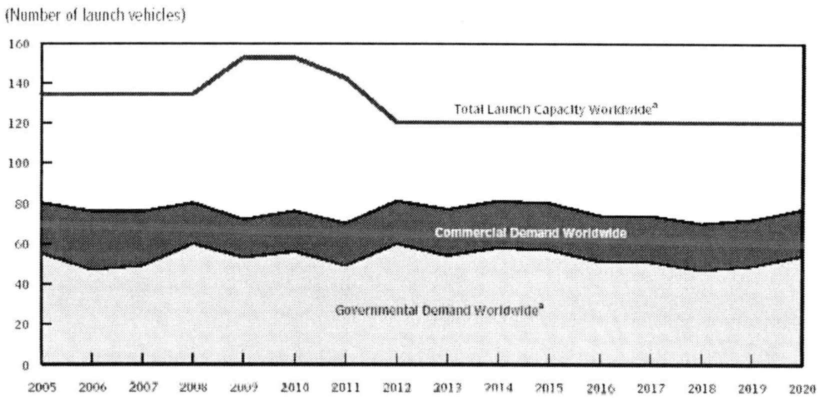

Source: Congressional Budget Office based on data provided by the Federal Aviation Administration and the Futron Corporation.

Note: Capacity is the number of launches that existing infrastructure and production facilities can support if fully funded and staffed. Demand is either the number of launches required on historical launch manifests or current projections of future launch manifests. For a variety of reasons, therefore (because of delays in the availability of payloads, for instance), the number of actual launches usually ends up being less than the demand reflected on manifests.

a. Excludes the National Aeronautics and Space Administration's Vision for Space Exploration initiative.

Figure 1.1. Worldwide Capacity and Demand for Launch Services

Box 1.1. Launch System Weight Classes and the Reference Orbit

The Federal Aviation Administration (FAA) distinguishes among launch vehicles on the basis of the maximum weight of payload that a given vehicle can place into orbit. The distinctions are as follows:

- Light = less than 2.3 metric tons (mt);
- Medium = 2.3 mt to 5.5 mt;
- Intermediate = 5.5 mt to 11.4 mt; and
- Heavy = greater than 11.4 mt.

The maximum payload that a launch vehicle can lift depends on both the launch site and the intended orbit. Therefore, to determine whether a given launcher is capable of lifting a light, medium, intermediate, or heavy payload, both the launch site and the orbit into which the payload is to be lifted must be specified. The FAA typically specifies a launch from Cape Canaveral, Fla., into a circular orbit that is 220 nautical miles (nm) in altitude at a 28.6 degree inclination to the equator. (A launch directly east from the Kennedy Space Center will result in an orbit inclined at the latitude of the center—28.6 degrees. The altitude of the orbit of the International Space Station, or ISS, is 220 nm.) Relative to that "reference" orbit, changes in altitude of plus or minus 100 nm would reduce or increase a given launch vehicle's maximum payload weight by about plus or minus 3 percent. Increasing the orbit's inclination to 56 degrees (the ISS's inclination) would reduce the launcher's maximum payload weight by about 8 percent.

If projections are restricted to payloads built by U.S. manufacturers and lifted into orbit using U.S. launch systems, supply exceeds demand through 2020, consistent with worldwide projections (see Figure 1.2).[17] Both U.S. supply and demand make up about 40 percent of the respective worldwide annual projections through 2010. By 2020, however, U.S. demand is projected to fall to about 35 percent of worldwide launch demand. That occurs because the on-orbit lifetime of U.S. satellites is assumed to continue to improve, requiring fewer launches of replacement satellites. Demand for governmental and commercial launches in Asia is assumed to increase as the economies of those countries, particularly China's and India's, continue to grow.

U.S. governmental demand for unmanned launches accounts for about 50 percent of total U.S. demand through 2020. That demand—about 94 percent of which is for payloads weighing less than 11.3 mt and none of which is for payloads exceeding 25 mt—is dominated by missions for the Department of Defense (DoD), NASA's unmanned earth and space science missions, and missions for the Department of Commerce, principally the launching of environmental-sensing satellites operated for weather forecasting and other purposes by the National Oceanic and Atmospheric Administration. DoD launches make up 63 percent of total governmental demand.

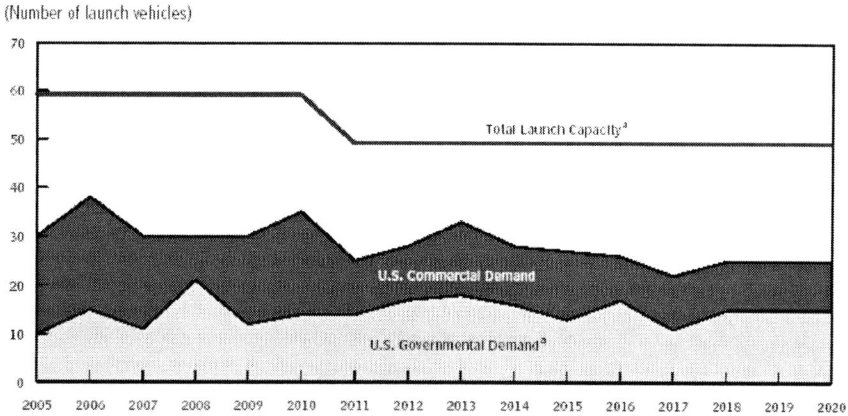

Source: Congressional Budget Office based on data provided in May 2005 by the Federal Aviation Administration, the Futron Corporation, the U.S. Air Force, and the National Aeronautics and Space Administration (NASA).

Note: Capacity is the number of launches that the infrastructure and production facilities can support if fully manned and funded. Demand is either the number of launches required on historical launch manifests or current projections of future launch manifests. For a variety of reasons, therefore (because of delays in the availability of payloads, for instance), the number of actual launches usually ends up being less than the demand reflected on manifests.

a. Excludes NASA's Vision for Space Exploration initiative.

Figure 1.2. Projections of U.S. Launch Capacity and Demand

Launch Requirements for the Vision for Space Exploration

The only demand currently projected for launches into orbit of payloads that weigh more than 25 metric tons applies to the manned lunar missions now planned as part of the VSE program. The treatment here briefly compares and contrasts manned lunar missions conducted under the Apollo program and those planned under the VSE. The discussion also considers future missions to Mars.

Lunar Missions Under the Apollo Program

The Saturn V launcher was capable of lifting a payload of somewhat less than 140 mt into LEO.[18] That payload consisted of the Apollo three-man command-and-service module (CSM), the ascent and descent stages of the lunar module (LM), the lunar excursion module (LEM), and the third stage of the Saturn V, which was used both to reach orbit and to propel the CSM, LM, and LEM out of low earth orbit and into a lunar trajectory. The propellant used to travel to the moon, descend to and ascend from the moon's surface, and return to Earth made up about 75 percent of the 140 mt payload.

Lunar Missions Under the VSE

According to current estimates, implementing a return to the moon as NASA now plans to do under the VSE program would require that a payload of about 150 mt be lifted into LEO. NASA plans to use propellants similar to those used for the Saturn/Apollo program and engines with similar efficiencies. Thus, about the same amount of propellant per pound of payload that was lifted into LEO during the Saturn/Apollo missions would be necessary for NASA's planned return to the moon. NASA's current plans for the VSE indicate that it is considering an increase in the size of the crew relative to that of Apollo (six for the VSE versus three for Apollo)—implying the need for a larger spacecraft. NASA is also considering lunar landings farther from the moon's equator and lunar missions that would last longer than a few days (as was typical during the Apollo program). Those changes would require more fuel for both the descent to and ascent from the moon and a greater mass of other consumables for life support. NASA is also planning to use the spacecraft that will transport humans to and from the moon—the CEV—to transport crew and limited amounts of cargo to and from the International Space Station after retirement of the space shuttle in 2010. However, NASA plans to use lightweight composites and state-of-the-art miniaturized electronics in constructing the VSE spacecraft, as well as improved techniques for insulating stored cryogenic fuel that are designed to reduce leakage during travel to and from the moon. NASA has estimated that the net effect of all those changes might increase the mass lifted into LEO for a return to the moon by approximately 10 percent to somewhat more than 150 mt.[19]

Although in the Apollo program the payload for lunar exploration was lifted into LEO with one launch, NASA considered other options for the VSE program, most of which would require multiple launches. The use of multiple launches would mean that less-capable and individually cheaper launchers could be used. In particular, such a strategy could permit the use of existing launchers rather than require the development of a new launcher with lift capability exceeding 100 mt. Multiple launches also would mean that a single launch failure would not necessarily cause overall mission failure. Further, multiple launches could permit the crew to be lifted on a launcher designed to be highly reliable, while less reliable launchers were used to lift other elements of the payload (cargo) that would fly to the moon.

There are, however, potential drawbacks associated with using multiple launches. Multiple launches require time to execute (perhaps several months), during which subcomponents of the full payload, some likely containing cryogenic fuels, must remain in LEO. Notwithstanding improved techniques for insulating cryogenic fuels, on- orbit storage of such fuels for many months would be a challenge and would require that fuel in excess of the amounts actually burned would need to be lifted, thereby potentially increasing the total number of launches needed.[20] The subcomponents would also have to be assembled into a complete spacecraft on-orbit, potentially requiring the development of a capability for reliable and fully automated rendezvous and docking, which the United States has never demonstrated. Multiple launches could also complicate mission planning because a launch failure or the failure of on-orbit rendezvous and docking would render a payload subcomponent unusable, and which subcomponent might fail could not be predicted a priori. Thus, additional payload subcomponents might need to be kept ready for launch in order to ensure completion of a successful mission.

As will be discussed in Chapter 2, the program alternatives considered by CBO for meeting the launch needs of manned lunar missions would require two to four launches. NASA is currently planning to use two launches into LEO to execute a single mission to the moon: one using the crew carrier to lift the crew and a small amount of cargo (which cumulatively weigh about 26 mt) and a second using the cargo carrier to lift the rest of the cargo (including the lunar lander and the fuel needed to travel to and from the moon, which weigh about 125 mt).[21] The reasons NASA cites for that choice include crew safety and the avoidance of complexities associated with on-orbit assembly in LEO of more than two payloads.[22]

Future Mars Missions

In nine NASA studies completed between 1988 and 2000, estimates of the weight of the payload that would need to be lifted into LEO for a Mars mission ranged between 470 mt and 1,500 mt. The variation depended on the propulsion system used for the long journey between LEO and Mars.[23] Therefore, the lightest Mars mission would require that a minimum of three to four times the mass of a lunar mission be launched into LEO.

NASA's plans for return missions to the moon avoid the complexities associated with on-orbit assembly that is dependent on more than two launches. However, such challenges for conducting a manned mission to Mars remain. Instead of developing a launch vehicle in at least the 400 mt class, which is not anticipated, capability for on-orbit assembly would be needed.[24] Assembling inert components in space can be either human-assisted or completely automated. If human-assisted, crew members could perform such operations on-orbit or control the assembly remotely. The United States has not demonstrated successful autonomous or remotely assisted assembly. Russia has demonstrated remotely assisted docking with its Progress resupply vehicle. NASA, however, has demonstrated only human-assisted, on-site assembly (for example, during construction of the International Space Station).

Unlike the mandated return to the moon, which has been slated for no later than 2020, the VSE states no specific schedule for a manned mission to Mars. Moreover, NASA's current planning through 2020 for manned space flight focuses on the funding and programs needed for returning humans to the moon. Consequently, this analysis, which examines launch needs through 2020, excludes consideration of launch capabilities and the associated technologies that may be needed to support human travel to Mars.

2. Alternatives for Future NASA Manned Space-Exploration Capabilities

The Congressional Budget Office considered two sets of alternatives for resuming manned space flight to the moon: those that would use existing heavy-lift launch vehicles and those that would require developing new launch systems. The new launch systems assessed by CBO include close derivatives of existing systems as well as designs that represent major departures from existing launch vehicles. The close-derivative launch vehicles would require relatively modest changes to the designs of existing launch systems and, in the case of cargo carriers, would result in vehicles that could lift several times the weight of current payloads. The vehicles that would constitute major departures from existing vehicles—referred to as super heavies—would be able to lift payloads weighing 100 metric tons or more.

Existing Heavy-Lift Launch Systems

Existing U.S. heavy-lift launch systems considered by CBO include the following:

- The National Aeronautics and Space Administration's space shuttle;
- Boeing's Delta IV Heavy variant of the U.S. Air Force's family of evolved expendable launch vehicles; and
- The most capable variant of Lockheed Martin's Atlas V Medium EELV.[25]

Payload is an important factor in distinguishing among those systems. The D4H has a lift capacity of about 25 metric tons, the most capable variant of the Atlas V has a lift capacity of 20.5 mt, and the space shuttle can lift a payload weighing 18 mt.[26] The Saturn V rocket that was used for the original manned Apollo missions to the moon carried a much heavier payload than existing systems do, lifting about 140 mt into low earth orbit with a single launch (see Figure 2.1). Therefore, using today's U.S. launch systems would require six to eight launches to lift into LEO all of the mass necessary to execute a manned lunar mission.

Challenges Using the Space Shuttle

The challenges associated with ensuring safe and reliable operation of the space shuttle have led NASA to decide to retire the shuttle fleet in about 2010, after the International Space Station is completed. Therefore, using the shuttle to execute manned lunar missions would require that decision to be reversed. Based on its historical performance, the shuttle would be expected to have a launch failure rate of about 2 percent.[27] Because at least eight shuttle launches would be required to deliver about 140 mt into LEO, the probability that all those launches would succeed is about 87 percent. Thus, there is a significant chance—13 percent—that at least one payload would fail to reach LEO.

As has been noted previously, which launch might fail could not be predicted. Moreover, once a failure occurred, experience indicates that a long delay would ensue before shuttle operations resumed because of concerns regarding crew safety. As a result, it is likely that the failure of any of the eight space shuttle launches needed to assemble the 150 mt payload required for a manned lunar mission would mean overall mission failure.

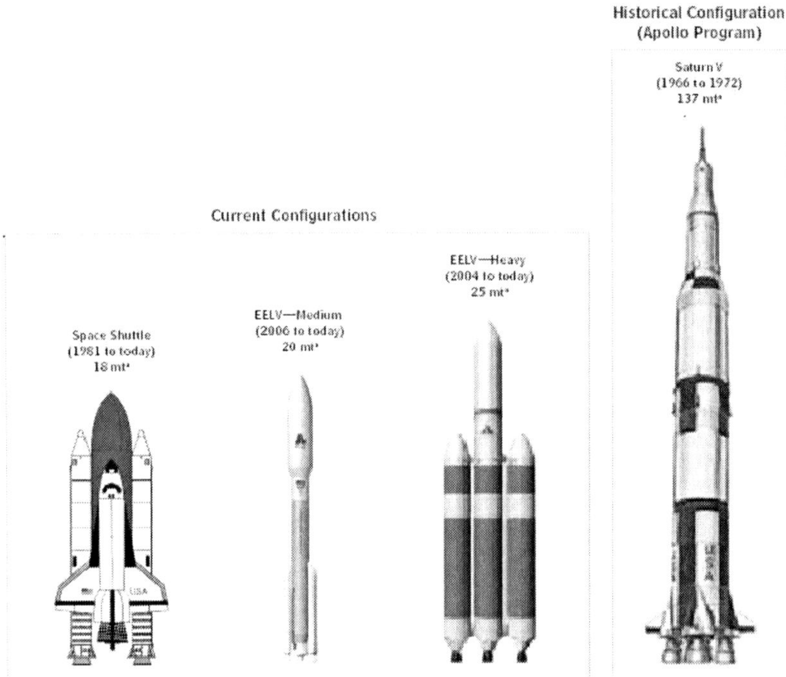

Source: Congressional Budget Office.

Note: Figures drawn to approximate scale; EELV = evolved expendable launch vehicle; mt = metric tons.

a. Metric tons of payload to a circular reference orbit of 220 nautical miles in altitude at a 28.6 degree inclination to the equator.

Figure 2.1. Existing and Historical U.S. Heavy-Lift Launch Vehicles

Even if all of the launches were successful, simply executing eight space shuttle launches in rapid succession would be a challenge. During the quarter-century that the shuttle program has been in operation, its launch rate has reached eight missions annually only four times. Currently, NASA hopes to execute two lunar missions annually, which would require an unprecedented 16 shuttle launches per year. Moreover, as indicated previously, the need to limit on-orbit leakage of cryogenic fuels would most likely require that the lunar mission be assembled in LEO over a maximum of about six months. That requirement suggests the need for monthly shuttle launch rates twice those that have been achieved historically to conduct a single lunar mission.

Table 2.1. Characteristics of Existing U.S. Heavy-Lift Launch Vehicles

Characteristic	Space Shuttle	Delta IV Heavy	Atlas V	Saturn V (1969)
Payload Capacity to LEO (Metric tons)	18	25	20	137
Operational Launch Success Rate (Percent)	98.28[a]	[c]	100	100
Number of Launches	116	1	1	13 [f]
Design, Development, and Testing (Billions of 2006 dollars)	~40	3.5[d]	2.0 [e]	31
Recurring Costs per Launch (Billions of 2006 dollars)	0.9	0.3	0.2	1.7
Recurring Costs per Pound of Payload (2006 dollars)	23,000/4,300[b]	5,400	4,500	5,500

Source: Congressional Budget Office.

Note: LEO = low earth orbit. (For all launch systems, this orbit is a circular reference orbit 220 nautical miles in altitude at a 28.6 degree inclination to the equator.)

a This statistic attributes the 2003 crash of the shuttle orbiter *Columbia* to launch failure.

b The first figure is the cost per pound of the payload. The second is the cost per pound if the orbiter, whose inert mass is about 78 metric tons, is considered part of the payload.

c The launch was not entirely successful because the test payload reached orbit but not the intended one.

d Includes contractor and government investment for the entire family of launchers, an entirely new RS-68 engine, and launch-pad modifications at the Cape Kennedy and Vandenberg launch sites.

e Includes contractor and government investment for the entire family of launchers and launch-pad modifications at the Cape Kennedy and Vandenberg launch sites.

f The 13 launches consist of two unmanned flights, one manned mission to LEO, nine lunar-mission launches, and one Skylab launch.

Finally, because the reusable space shuttle returns to Earth and has to be refurbished before each launch, it is relatively expensive to use. Therefore, using the shuttle to lift a pound of payload into orbit costs about four times more than did using the Saturn V (see Table 2.1).

Challenges Using the Delta IV Heavy EELV

The launch reliability of systems such as the Delta IV class of vehicles should be similar to that of their Delta predecessors, as well as to that of the space shuttle (see Appendix A).[28] The number of launches required for the D4H to lift about 150 mt into LEO would be six, implying about a 10 percent chance that one payload would fail to reach orbit. Therefore, the use of the D4H to return humans to the moon would raise issues regarding overall mission success similar to those associated with using the space shuttle.

The Delta IV Heavy launch system was not designed for human space flight, however, which creates problems in complying with the safety standards that NASA has established for manned space flight. (At least one of the payloads lifted into LEO for a manned lunar mission would consist of the spacecraft that would travel to the moon with its crew.) Those problems include:

- A lack of redundant systems;
- Insufficient safety margins in the strength of the structures that compose the launcher;
- A lack of sensors and systems for monitoring the performance of the launch vehicle during flight—called health monitoring equipment—which could alert the crew to an impending failure and the need to execute a "launch escape"; and
- A need for modifications to the Delta launch pad to enable a crew to escape, if necessary, prior to launch.

A particular issue regarding the safety of manned space flight that is associated with the Delta IV Heavy launch vehicle is the trajectory it must follow as it lifts humans into orbit. The trajectory that enables the D4H to lift its full 25 mt payload tends to be especially steep, allowing the vehicle to ascend directly to high altitude. If the launcher developed a problem during ascent that required the crew to perform a launch escape by separating the crew from the booster, the ballistic trajectory on which the crew would descend to Earth would be similarly steep. The deceleration experienced by the crew during that steep descent into the atmosphere would exceed the safety limits specified by NASA, which are established to ensure that the crew does not suffer serious injury (see Appendix B). Therefore, using the D4H would require either that NASA's safety standards be relaxed or that the launcher fly a less direct trajectory. Use of a more gradual, or so-called depressed, trajectory would require more fuel to be expended per pound of payload lifted, reducing to under 20 mt the maximum weight of payload that the D4H could lift.

Challenges Using the Atlas V Medium EELV

The most capable version of the Atlas V Medium that is currently available can lift a 20.5 mt payload into LEO, which would require at least seven launches to execute a manned lunar mission. Therefore, its use to return humans to the moon would raise issues regarding overall mission success similar to those posed by either the space shuttle or the Delta IV Heavy. With some reduction in payload mass, the trajectory that the Atlas V followed could also be tailored to avoid the potential for crew injury during a launch abort. However, using the Atlas V would require resolution of the same problems related to redundancy, safety margin, health monitoring, and crew escape that are associated with the Delta IV Heavy launcher. Use of the A5M would further require that its Russian-made RD-180 engine be certified as acceptable for manned space flight. According to NASA staff, such certification would require the development of the capability for a U.S. supplier to manufacture and test the RD-180.

Close Derivatives of Existing Systems

CBO characterizes as "close derivatives" those new launch vehicles that retain a close pedigree to existing systems. Although those close derivatives may have the capacity to lift payloads two to four times heavier than existing launch vehicles, that increase in payload weight could be achieved by making major modifications to just a few components of an existing launch system. In this study, CBO analyzes the cost and performance of close- derivative launch vehicles that could be used to execute a manned lunar mission. The vehicles are similar to those considered by NASA and members of the U.S. space-launch industry, including Boeing, Lockheed Martin, and ATK Thiokol.[29] The close derivatives considered by CBO include modified versions of the space shuttle and its components, as well as modified versions of both the Atlas and Delta versions of the EELV. In all cases, the launchers used to lift the spacecraft and crew (called crew carriers) into LEO are distinct from the launchers (called cargo carriers) used to lift the remaining payload, such as the lunar lander and fuel needed to support flight to and from the moon.

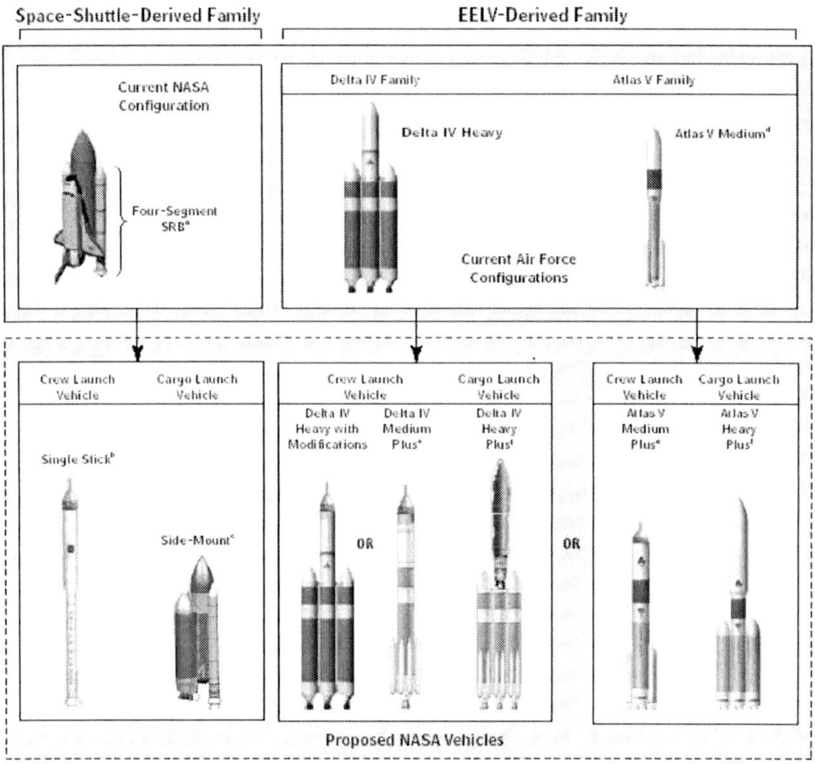

Source: Congressional Budget Office.

Note: Figures drawn to approximate scale; EELV = evolved expendable launch vehicle; SRB = solid rocket booster; NASA = National Aeronautics and Space Administration. The term "close derivative" describes those new launch vehicles that retain a close pedigree to existing systems. Figures may be family representatives.

a. The main propulsion system for the shuttle is composed of three "sticks": two four-segment SRBs attached to the central external tank (the third stick).

b. A single five-segment stick based on a four-segment SRB.

c. With cargo canister on side.

d. The A5 551 is the largest of the Atlas V medium-lift launchers. A design for a triple-stick Atlas V heavy-lift vehicle has completed Air Force review, but no production is planned.

e. The "Medium Plus" refers to an upgraded version of the medium (single-stick) branch of the respective EELV family of launchers.

f. The "Heavy Plus" refers to the triple-stick version of the medium plus branch of the respective EELV family of launchers.

Figure 2.2. Close-Derivative Launchers Considered by CBO and Their Antecedents

Table 2.2. Selected Characteristics of the Close-Derivative Launchers Considered by CBO

Alternative	Primary Mission	Payload to LEO (mt)	Launch-Abort Acceleration[b]
		Shuttle-Derived	
Five-Segment Stick[a]	Crew Carrier	26	10.6
Side-Mount	Cargo Carrier	77	c
		EELV-Derived	
Delta IV Heavy with Modifications (D4H with Mods)	Crew Carrier	21	18
Delta IV Medium Plus (D4M +)	Crew Carrier	18	18
Delta IV Heavy Plus (D4H+)	Cargo Carrier	40	c
Atlas V Medium Plus (A5M +)	Crew Carrier	24	16.5
Atlas V Heavy Plus (A5H+)	Cargo Carrier	74	c

Source: Congressional Budget Office based on data provided by the National Aeronautics and Space Administration, Boeing, Lockheed Martin, and ATK Thiokol.

Note: EELV = evolved expendable launch vehicle; LEO = low earth orbit (for all launch systems, this orbit is a circular reference orbit 220 nautical miles in altitude at a 28.6 degree inclination to the equator); mt = metric tons.

a. The main propulsion system for the shuttle is composed of three "sticks": two four-segment solid rocket boosters attached to the central external tank (the third stick).

b. If a launch fails, the maximum acceleration—expressed in multiples of gravity's acceleration—that a crew could experience in uncon¬trolled ballistic reentry.

c. Not a relevant characteristic for a cargo carrier.

Shuttle-Derived Launch Vehicles

- **The Five-Segment "Stick" (crew carrier/lift capacity, 26 mt).** This launch vehicle is based on the four- segment solid rocket booster, a pair of which now help lift the space shuttle into orbit (see Figure 2.2). The main propulsion system for the shuttle is composed of three "sticks": two four-segment SRBs attached to the central external tank, or ET (the third stick). The ET carries the liquid hydrogen and liquid oxygen that fuel the space shuttle main engines housed in the rear portion of the orbiter (which is also attached to the ET). The SRBs are composed of individual cylindrical sections, each containing a casting of solid fuel. The single five-segment stick considered here would consist of five such sections, all of which would constitute the first stage of the rocket. A new second stage (with an independent liquid-fuel propulsion system) would be mounted on top of the SRB. The additional stage is needed to provide the capability to lift into LEO 26 mt. (NASA refers to that spacecraft as the crew exploration vehicle, or CEV.) As of September 2005, NASA's proposed solution for the crew carrier was a four-segment stick with a second stage propelled by modified versions of the space shuttle main engines. However, with the submission to the Congress of the President's fiscal year 2007 budget request, NASA changed that proposal to a five-segment stick using a second stage equipped with a variant of the less powerful J-2S engine—a modified version of the J-2 engine used on the second and third stages of the Saturn V. That combination of engines and booster would require that the engines in the CEV provide propulsion to attain the needed circular low earth "parking" orbit.

- **The Side-Mount (cargo carrier/lift capacity, 77 mt).** The Side-Mount would essentially consist of the space shuttle propulsion system with a cargo canister replacing the airplane-like orbiter. (Because it must house and protect a human crew and carry up to 18 mt of cargo to and from LEO, the 80-ton orbiter is relatively massive. Its structure, including its wings, must also withstand the heat and force of atmospheric reentry.) The cargo canister would be relatively light because it would be expendable. Therefore, the payload that the Side-Mount could lift into LEO would increase to about 77 mt (see Table 2.2).

EELV-Derived Launch Vehicles

- **The Delta IV Heavy with Modifications, or D4H with Mods (crew carrier/lift capacity, 21 mt).** The modifications that CBO considered would provide the component redundancy, structural safety margins, and health monitoring equipment required for manned flight. The D4H with mods, like the existing D4H, would be a three-stick launch vehicle with about three times the number of first-stage hardware components featured on the less capable, single-stick versions of the Delta IV. That additional hardware, however, also introduces additional opportunities for failure, making the D4H with mods less reliable than the shuttle-derived, single-stick launchers discussed previously.[30] The D4H with mods would be capable of lifting about 21 mt into a circular LEO.

- **The Delta IV Medium Plus, or D4M+ (crew carrier/lift capacity, 18 mt).** Relative to the single-stick Delta IV Medium (D4M) EELV—which was developed for the U.S. Air Force and has executed several successful launches—the D4M+ would feature an enhanced first-stage core with six solid rocket boosters and an expanded second stage with additional engines. Those modifications would enable the D4M+ to lift about 18 mt into a circular LEO flying a depressed trajectory. Designed as a single-stick launcher, it would be more reliable than the D4H with mods.

- **The Delta IV Heavy Plus, or D4H+ (cargo carrier/lift capacity, 40 mt).** The Delta IV Heavy is built using three of the first-stage central cores featured on the Delta IV Medium. Similarly, the D4H+ would use three of the central cores of the D4M+, enabling it to lift roughly 40 mt of cargo into LEO.

- **The Atlas V Medium Plus, or A5M+ (crew carrier/lift capacity, 24 mt).** Developed for the Air Force, the Atlas V Medium has carried out several successful launches. Relative to the existing A5M launcher, the A5M+ would have a larger first-stage central core and an upgraded second stage providing the capability to lift about 20 mt into LEO. In addition, the A5M+ would incorporate the structural safety margins, component redundancy, and health monitoring equipment needed for manned space flight. Although the most capable version of the existing A5M can also lift 20 mt, it requires the use of five strap-on SRBs (attached to the first-stage central core) to do so. Because of its expanded first stage, the A5M+ would not need to use strap-on boosters, and its first stage would serve as the building block for a heavy-lift cargo carrier. The A5M+ would be capable of lifting about 24 mt into a circular LEO.

Table 2.3. Comparison of the Costs of the Close-Derivative Launchers Considered by CBO

(Billions of 2006 dollars)			
Alternative	**Primary Mission**	**Nonrecurring Development Cost**	**Unit Recurring Cost (Five launchers per year)**
		Shuttle-Derived	
Five-Segment Stick[a]	Crew Carrier	4.8	0.45
Side-Mount	Cargo Carrier	4.7	0.95
		EELV-Derived	
Delta IV Heavy with Modifications (D4H with Mods)	Crew Carrier	2.9	0.47
Delta IV Medium Plus (D4M+)	Crew Carrier	3.3	0.40
Delta IV Heavy Plus (D4H+)	Cargo Carrier	5.2	0.60
Atlas V Medium Plus (A5M +)	Crew Carrier	5.3	0.40
Atlas V Heavy Plus (A5H+)	Cargo Carrier	4.0	0.60

Source: Congressional Budget Office.

Note: EELV = evolved expendable launch vehicle.

a. The main propulsion system for the shuttle is composed of three "sticks": two four-segment solid rocket boosters attached to the central external tank (the third stick).

- **The Atlas V Heavy Plus, or A5H+ (cargo carrier/lift capacity, 74 mt).** The first stage of the Atlas V Medium Plus would serve as the basis for this three-stick vehicle. Relative to the A5M+, the A5H+ would feature additional engines in its second stage. It would have the capability to lift about 74 mt into LEO.

Comparisons of the Costs of the Close-Derivative Launchers

CBO's estimates for the costs of the launchers described above indicate the following (see Table 2.3):

- On a per-vehicle basis, recurring costs for the shuttle-derived crew carrier and the EELV-derived crew carriers would be about the same, within 10 percent.
- On a per-vehicle basis, recurring costs for the EELV-derived cargo carriers would be about one-third less than those for the shuttle-derived Side-Mount. Those lower costs reflect, in part, the potential

for the sharing of fixed costs (such as for launch pads and production facilities) that is possible between EELV-derived vehicles and DoD's EELV program.[31] Such sharing is not possible with shuttle-derived vehicles because only NASA uses them. EELV-derived cargo carriers also have less lift capacity than the shuttle-derived cargo carrier: reduced performance implies lower costs for each individual launch vehicle. But, in the case of the EELV-derived D4H+, that reduced performance also means that more cargo carriers would have to be launched to conduct a lunar mission.

- Nonrecurring costs for the design, development, and testing of the EELV-derived launchers would be about the same, on average, as those for shuttle-derived launchers.

The methodology and assumptions on which these results are based are discussed in detail in Appendix E.

New Super-Heavy Launch Vehicles

The super-heavy launch vehicles considered by CBO are designed to lift payloads of 100 mt or more. In general, those prospective launchers would require physically significant and relatively expensive upgrades to many—and, in some cases, nearly all—of the major systems that make up existing launchers. Thus, developing super-heavy launchers would more than likely entail greater risks of cost growth and schedule slippage than would be the case with the close-derivative launchers discussed previously. A primary consideration would be the need to accommodate a larger number of engines in the first stage and, when present, the second stage.[32] (For depictions of the super-heavy launchers considered by CBO and comparisons of those vehicles to closely related existing launchers, see Figure 2.3. For greater detail, see Appendix D.) CBO considers super-heavy launch vehicles that use components of the space shuttle, as well as launchers that would represent new and much more capable designs that incorporate selected components (primarily engines) used on the current EELVs. CBO does not consider as an alternative rebuilding the Saturn V. Duplicating the infrastructure and engineering expertise needed to rebuild and reconfigure the Saturn V would probably make such an alternative unattractive when compared with alternatives that rely, at least in part, on the infrastructure and expertise associated with currently operational launchers.[33]

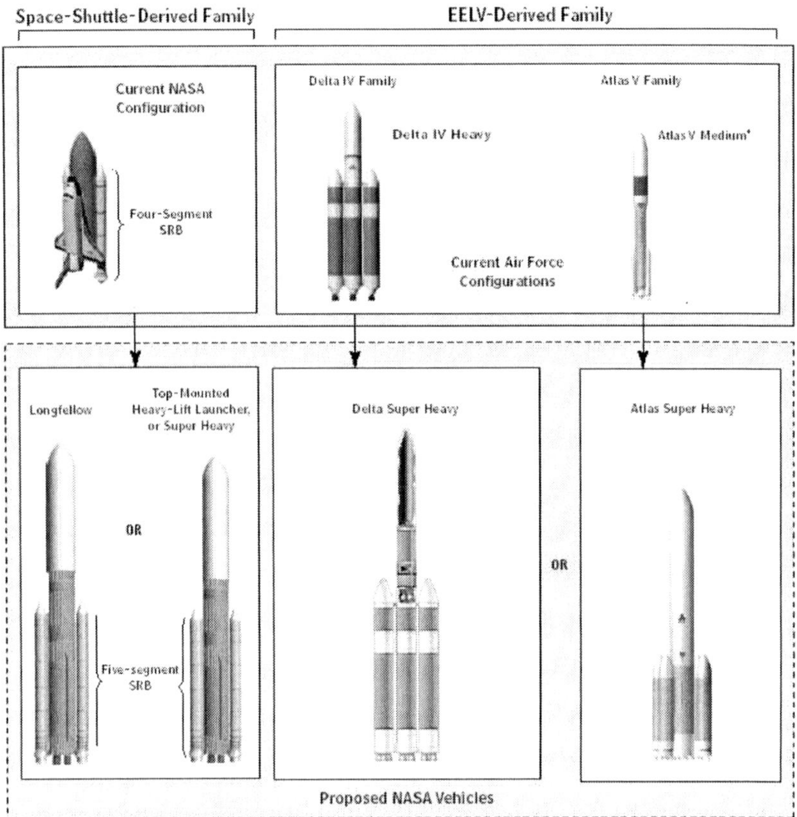

Source: Congressional Budget Office.

Note: Figures drawn to approximate scale; EELV = evolved expendable launch vehicle; NASA = National Aeronautics and Space Administration; SRB = solid rocket booster.

a The A5 551 is the largest of the Atlas V medium-lift launchers. A design for a triple-stick Atlas V heavy-lift vehicle has completed Air Force review, but no production is planned.

Figure 2.3.Super-Heavy Launchers Considered by CBO as Cargo Launch Vehicles and Their Antecedents

Super-Heavy Launchers That Use Shuttle Components

- **The Top-Mounted, Shuttle-Derived Super Heavy (cargo carrier/lift capacity, 125 mt).** In the President's budget for 2007, this cargo launch vehicle was identified as NASA's recommended solution for supporting manned lunar missions.[34] Because many of its major components would be drawn from the space shuttle, CBO refers to the vehicle as the shuttle-derived super heavy. Designed to be about the size of the Saturn V rocket, the vehicle would be able to lift a payload to LEO weighing about 125 mt (see Table 2.4). It would use five-segment solid rocket boosters (as opposed to the four-segment SRBs used on the shuttle). Two of those five-segment SRBs would be mounted on the sides of an elongated version of the shuttle's existing external tank, which would be modified to house five expendable versions of the existing (reusable) SSMEs. Because the vehicle has no second stage, the payload engine designated to continue powering the assembled lunar mission from low earth orbit would need to be burned twice: once to establish the needed circular low earth parking orbit; and a second, subsequent time to propel the assembled mission on a trajectory to the moon. A large cargo canister or shroud would be mounted on top of the ET/SRB assembly to house the payload.

- **The Longfellow (cargo carrier/lift capacity, 108 mt).** This launcher—proposed by an industry consortium that includes Boeing, Lockheed Martin, and ATK Thiokol—would resemble the shuttle-derived super heavy but with several different design aspects. Rather than expendable versions of SSMEs, the first stage of this launcher would rely on four of the RS-68 engines currently used on the Delta IV EELVs. It would also have a second stage powered by a single J-2S engine, atop which would be mounted a cargo canister. Because the Longfellow would be able to lift about 108 mt into LEO, overall its four RS-68 engines and single J-2S engine would provide less capability than the five SSMEs used on the shuttle-derived super heavy (see Table 2.4).[35]

Super-Heavy Launchers with EELV Antecedents

- **The Delta Super Heavy (cargo carrier/lift capacity, 146 mt).** The closest antecedent of this vehicle—proposed by Boeing—would be the existing Delta IV Heavy. The new launch vehicle, however, would incorporate many changes relative to the prospective Delta IV Heavy Plus. Those changes include the use of three expanded first-stage cores, each 8 meters in diameter and each equipped with four RS-68 engines. Thus, the three-core first stage of this launch vehicle would have 12 engines rather than the three-core, three-engine first stage of the existing D4H. The Delta super heavy would also have a larger second stage (8 meters in diameter) equipped with a single SSME. That SSME would be modified to start at high altitude (in contrast to the existing SSME, which is designed to start at sea level). In terms of physical dimensions, this launch vehicle would be the largest of the super heavies considered by CBO. It also would be capable of lifting the largest payload—about 146 mt—into LEO (see Table 2.4).

Table 2.4. Selected Characteristics of the Super-Heavy Launchers Considered by CBO

Alternative	Primary Mission	Payload to LEO (mt)
Shuttle-Derived		
Longfellow	Cargo Carrier	108
Super Heavy	Cargo Carrier	125
EELV Antecedent		
Delta Super Heavy	Cargo Carrier	146
Atlas Super Heavy	Cargo Carrier	135

Source: Congressional Budget Office based on data provided by the National Aeronautics and Space Administration, Boeing, Lockheed Martin, and ATK Thiokol.

Note: EELV = evolved expendable launch vehicle; LEO = low earth orbit (for all launch systems, this orbit is a circular reference orbit 220 nautical miles in altitude at a 28.6 degree inclination to the equator); mt = metric tons.

Table 2.5. Comparison of the Costs of the Super-Heavy Launchers Considered by CBO

(Billions of 2006 dollars)			
Alternative	Nonrecurring Development Cost	Unit Recurring Cost (Two launchers per year)	Recurring Cost per Pound of Payload (Dollars per pound)
		Shuttle-Derived	
Longfellow	8.0	1.1	4,600
Super Heavy	8.9	1.3	4,700
		EELV Antecedent	
Delta Super Heavy	16.7	1.5	4,700
Atlas Super Heavy	9.0	1.2	4,100

Source: Congressional Budget Office.

Note: EELV = evolved expendable launch vehicle.

- **The Atlas Super Heavy (cargo carrier/lift capacity, 135 mt).** The closest antecedent of this vehicle—proposed by Lockheed Martin—would be the Atlas V Heavy Plus. Relative to that prospective launcher, the Atlas super heavy would have a substantially expanded 8.4-meter first-stage central core powered by five RD-180 engines (as opposed to the two that power the A5H+). The expanded central core would be surrounded by four of the smaller first-stage cores used in the A5H+'s triple-stick configuration. Thus, the first stage of the Atlas super heavy would consist of a five-stick configuration with 13 engines (see Figure 2.3). The launcher's second stage would be similar to that of the A5H+ but would incorporate a longer payload shroud. Those changes would result in a vehicle capable of lifting 135 mt into LEO (see Table 2.4). In physical dimensions, the Atlas super heavy would not need to be as large as the other super-heavy launchers considered by CBO. That difference is due primarily to the type of liquid fuel used for its first-stage engines: unlike the other prospective super-heavy launchers, which would use liquid oxygen and liquid hydrogen, the Atlas super heavy would consume liquid oxygen and a kerosene variant. The use of less-dense liquid hydrogen requires larger tanks, necessitating larger first-stage assemblages for the prospective super-heavy launchers that use space shuttle or Delta IV components.

Comparisons of the Costs of Super-Heavy Launchers

The recurring launch costs of the super-heavy launchers considered by CBO range from $1.1 billion to $1.5 billion; their nonrecurring development costs range from $8 billion to nearly $17 billion (see Table 2.5).[36] By comparison, historical data indicate that the recurring launch costs of the Saturn V were about $1.7 billion and that its nonrecurring development costs were about $30 billion. The Saturn V's greater nonrecurring costs were attributable to the development of its engines and other associated technology—work that would not have to be repeated for the super-heavy launchers considered by CBO.

Unlike the case for close-derivative cargo launchers, the super-heavy launchers based on shuttle components are not necessarily more expensive than super-heavy launchers with EELV antecedents. The Atlas and Delta super-heavy launchers would be much larger than their EELV antecedents. Therefore, they could not as easily share the launch pads, manufacturing facilities, ground-support equipment, and other support facilities that the prospective close-derivative Atlas and Delta launchers would share with existing EELVs. The absence of the opportunity to share infrastructure is the primary reason that the Longfellow, Atlas, and shuttle-derived super-heavy launchers all have recurring and nonrecurring costs that are relatively similar.

Except for the Delta super heavy, nonrecurring costs for the super-heavy launchers would amount to about $9 billion. The Delta super heavy's nonrecurring costs would be significantly higher, primarily because of its size (and larger payload). Building and assembling the large-volume, triple-stick configuration of the Delta super heavy would require substantial modifications to existing manufacturing, integration, fuel storage and handling, and launch facilities—even if the shuttle launch pad was used for the Delta super heavy.[37] Accommodating those needs would cause nonrecurring costs for this launcher to be higher than corresponding costs for the other super-heavy launchers considered by CBO, which would make greater use of existing facilities that required fewer modifications (see Appendix E).

The Longfellow lifts the smallest payload into LEO of any of the prospective super-heavy launchers—108 mt. Thus, two Longfellow launchers would be needed to lift the cargo associated with a single manned lunar mission, basically negating its lower per-launch cost.[38]

According to CBO's estimates, the unit cost of the Atlas super heavy would be 10 percent less than that of the shuttle-derived super heavy.[39] Nonetheless, some observers might argue that the shuttle-derived launcher would be more cost-effective than the Atlas. Relative to the shuttle-derived super heavy, the first stage of the Atlas super heavy would be more complex because of its larger number of engines (15 versus seven) and sticks (five versus three). Generally, greater complexity implies reduced reliability (because of greater opportunities for potential failure).[40,41] If the Atlas failed more often than the shuttle- derived super heavy, its use might be more expensive overall because of the need for additional launches to compensate for failures.

Program Cost Comparisons

In this section, CBO provides projections for the funding that NASA would need for launch vehicle development and procurement to execute manned lunar missions using the alternative crew and cargo launchers described previously. Those CBO projections are compared with NASA's current projection of the funding it estimates will be needed through 2017 to develop and buy launch vehicles capable of supporting both flight to the International Space Station and manned lunar missions.

To generate those projections, CBO needed to make a variety of "programmatic" assumptions, such as when the vehicles would first be used to launch either cargo or a spacecraft and its crew and how many test flights would be necessary before a launch vehicle was deemed ready for use. CBO used two sources for building those assumptions. First, to the extent that the September 2005 Exploration Systems Architecture Study documented the programmatic assumptions NASA is using, CBO's projections are built on those assumptions—unless they have been subsequently superseded by materials the agency prepared for the President's budget. In particular, justification materials provided for the 2007 budget indicate that NASA remains hopeful it can meet the schedule outlined in the ESAS for conducting the first flight of the CEV in 2012 and the first manned lunar mission in 2018. However, the agency recognizes that budget constraints might delay those events until 2014 and 2020, respectively. Therefore, CBO has constructed budgetary projections for launch vehicle funding needed to execute both of those schedules. The programmatic assumptions that CBO made include the following:

- Certification of a launch vehicle—whether a crew carrier or a cargo carrier—for operational use requires three test or demonstration (test/demo) flights.
- Appropriations for hardware production and launch services precede a launch by two years.
- The first operational flight of the CEV (carried aloft by the crew carrier) occurs in either 2012 (the more ambitious schedule) or 2014 (the less ambitious schedule). The first manned lunar mission occurs in either 2018 (the more ambitious schedule) or 2020 (the less ambitious schedule).
- Five ISS support flights per year are conducted through 2016 beginning in either 2012 or 2014.[42] Regardless of whether the payload to the ISS includes a crew, the launcher for those flights will be the crew launch vehicle. With the retirement of the space shuttle projected to occur in 2010, this assumption implies a hiatus in manned space flight of two to four years.
- Two manned lunar missions occur annually, beginning in either 2018 or 2020.
- No budgetary allowances are made for launch failure.

Table 2.6. Cargo Launches Needed to Execute a Manned Lunar Mission with the Alternative Cargo Launchers Considered by CBO

Alternative	Payload to LEO (mt)	Cargo Launches per Lunar Mission	Maximum Cargo Payload per Lunar Mission (mt)[a]
		Shuttle-Derived	
Side-Mount	77	2	154
Longfellow	108	2	216
Super Heavy	125	1	125
		EELV Antecedent	
Delta IV Heavy Plus (D4H+)	40	3	120
Atlas V Heavy Plus (A5H+)	74	2	148
Delta Super Heavy	146	1	146
Atlas Super Heavy	135	1	135

Source: Congressional Budget Office based on data provided by Boeing, Lockheed Martin, and ATK Thiokol.

Note: EELV = evolved expendable launch vehicle; LEO = low earth orbit (for all launch systems, this orbit is a circular reference orbit 220 nautical miles in altitude at a 28.6 degree inclination to the equator); mt = metric tons.

a. Does not include 18 to 28 mt delivered to LEO by a separate launch of the crew carrier.

The more ambitious schedule would require that 29 crew carriers be purchased to conduct operational launches through 2017.[43] Twenty-five of those vehicles would support ISS missions, and four would be designated for lunar missions. Executing the more ambitious schedule would require purchasing four to 12 cargo carriers through 2017 to conduct lunar missions starting in 2018. The range in the number of needed cargo carriers reflects the differing number of launches associated with each of the alternatives that CBO considered (see Table 2.6).

The less ambitious schedule implies that 15 crew carriers would be purchased through 2017 to conduct operational launches. All of those launchers would be used for ISS support missions. Executing the less ambitious schedule would not require cargo carriers to be purchased through 2017 to conduct operational launches, however.

The Vision for Space Exploration Initiative: Projected Funding for Launch Vehicles

Through 2017, NASA has projected the funding that it anticipates will be needed to develop and purchase launch vehicles to conduct missions to the ISS and to the moon. Four accounts are directly related to VSE launch vehicles under what the agency calls its Constellation Systems Program, as follows:

- Crew Launch Vehicle. This account also supports the delivery of crew and cargo to the ISS using the five- segment launcher.
- Heavy-Lift Launch Vehicle (the cargo carrier).
- Launch and Mission Systems (L&MS). This account covers processing, launch operations, and research and development of facilities, ground equipment, control systems, and communications networks for the crew launch vehicle, the cargo carrier, and the CEV. NASA budget materials available to CBO do not specify how much L&MS funding will be allocated to the CEV and how much to the launchers. Therefore, CBO assumes that, through 2010, L&MS funding is solely associated with the CEV. After 2010, two-thirds of the funding is assumed to be associated with crew and cargo launchers.

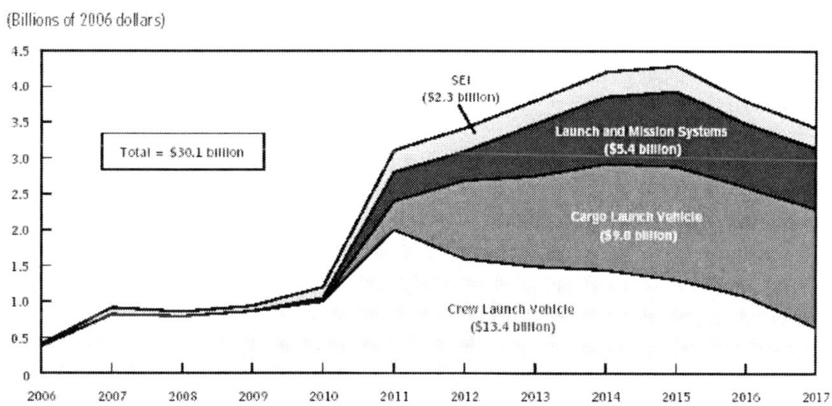

Source: Congressional Budget Office based on the President's budget for fiscal year 2007.

Note: NASA = National Aeronautics and Space Administration; SEI = system engineering and integration; VSE = Vision for Space Exploration.

Figure 2.4. NASA's Projected Funding for VSE Launch Vehicles Through 2017

- System Engineering and Integration (SEI). This account funds activities that support all aspects of the Constellation Systems Program. The NASA budget-materials available to CBO do not specify how much SEI funding will be allocated to support development and operational launches of the crew and cargo carriers. Therefore, CBO assumes that the fraction of SEI funding allocated to support launch vehicles will be equal to the fraction of funding for the Constellation Systems Program (less SEI) composed of the three launch vehicle accounts listed immediately above.

Through 2017, NASA's projections for cumulative funding in those four accounts, which is slated to support the development and purchase of VSE launch vehicles, total $30.1 billion. Through 2010, the projected amounts are about $1.0 billion annually. After 2010, annual amounts are about $3.5 billion (see Figure 2.4). NASA's projection is constructed to support the development and use of the five-segment stick as the crew carrier and the shuttle-derived super-heavy launcher as the cargo carrier.

Comparisons of Program Costs Using Close-Derivative Launchers

CBO has constructed budgetary projections through 2017 for three alternative launch programs that would use close-derivative launchers, as follows:

- **The "pure Atlas-derived" alternative.** This program would use the A5M+ launcher as the crew carrier and the A5H+ launcher as the cargo carrier.

- **The "pure Delta-derived" alternative.** This program would feature the D4M+ with mods as the crew carrier and the D4H+ as the cargo carrier.

- **The "pure shuttle-derived" alternative.** This program would use the five-segment stick as the crew carrier and the Side-Mount as the cargo carrier.

The above three alternative programs do not constitute a complete set of alternatives that could be pursued using the four close-derivative crew carriers and three close-derivative cargo carriers described previously. CBO has chosen to analyze this subset in order to limit the number of budgetary projections it built to a manageable number while still illustrating the differences in costs associated with pursuing EELV-derived or shuttle-derived programs.

To execute the more ambitious schedule for returning humans to the moon, CBO estimates that total costs through 2017 for pursuing the three alternatives it considered using close-derivative launchers would range from about $26 billion for the pure Atlas-derived launch program to almost $31 billion for the pure shuttle-derived program (see Figure 2.5). Executing the less ambitious schedule would cost $15 billion for the pure Delta-derived program to $16 billion for the pure shuttle- derived program through 2017, CBO estimates.

Comparisons of Program Costs Using Super-Heavy Launchers

CBO has constructed budgetary projections through 2017 for three alternative launch programs that would use new super-heavy launchers, as follows:

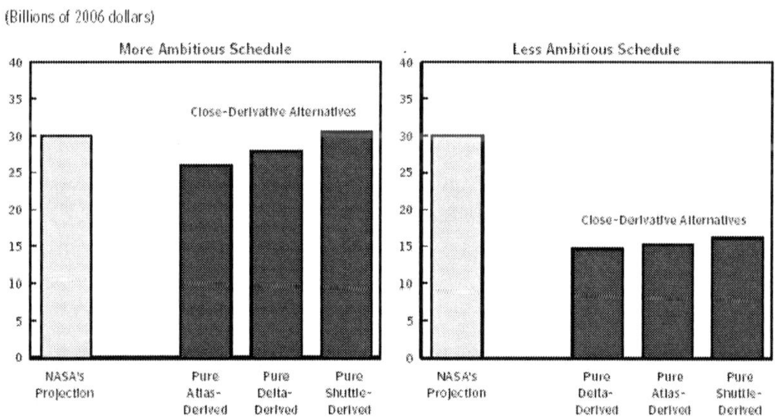

Source: Congressional Budget Office.

Notes: NASA = National Aeronautics and Space Administration.

The more ambitious schedule initiates International Space Station support in 2012 and the first lunar mission in 2018. The less ambitious schedule initiates International Space Station support in 2014 and the first lunar mission in 2020.

The term "pure" means that both the crew carrier and the cargo carrier are derived from vehicles in the same family of launchers. The term "close derivative" describes new launchers that retain a close pedigree to existing systems.

Figure 2.5. Costs Through 2017 of Using Close-Derivative Launchers to Conduct Manned Lunar Missions

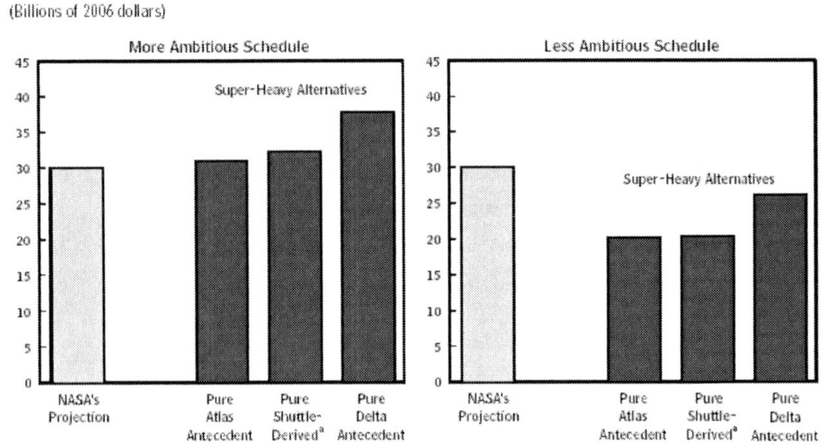

Source: Congressional Budget Office.

Notes: NASA = National Aeronautics and Space Administration.

The more ambitious schedule initiates International Space Station support in 2012 and the first lunar mission in 2018. The less ambitious schedule initiates International Space Station support in 2014 and the first lunar mission in 2020.

Each alternative includes the close-derivative crew carrier associated with the indicated family. The term "pure" means that both the crew carrier and the cargo carrier are derived from vehicles in the same family of launchers.

The term "close derivative" describes new launchers that retain a close pedigree to existing systems.

a. NASA's choice of crew launch vehicle and cargo carrier as outlined in the President's budget for 2007.

Figure 2.6. Costs Through 2017 of Using Super-Heavy Launchers to Conduct Manned Lunar Missions

- **The "pure Atlas-antecedent" alternative.** This program would use the A5M+ launcher as the crew carrier and the Atlas super-heavy launcher as the cargo carrier.

- **The "pure Delta-antecedent" alternative.** This program would use the D4M+ as the crew carrier and the Delta super-heavy launcher as the cargo carrier.

- **The "shuttle-derived super-heavy" alternative.** This program would use the five-segment stick as the crew carrier and the top-mounted, shuttle-derived super-heavy launcher as the cargo carrier.

Upon submission of the 2007 budget, this was NASA's choice for VSE launch vehicles.[44]

To implement the more ambitious schedule using super-heavy launchers, CBO estimates that costs through 2017 would range from $31 billion to $38 billion (see Figure 2.6). CBO's estimates indicate that using NASA's choice of launch vehicles to execute the more ambitious schedule would cost about $2 billion more than the funding contained in NASA's budget projection through 2017.[45]

Box 2.1. Risks Associated with Developing and Producing NASA's Launch Vehicle Choices

The National Aeronautics and Space Administration (NASA) has acknowledged that its choice of launch vehicles, particularly the five-segment stick proposed by the agency as a crew carrier, presents a number of risks.[1] Such risks could cause actual costs to exceed, perhaps substantially, estimates made by the Congressional Budget Office (CBO). The risks cited by NASA personnel relate to the following functions:

- **Stability control.** A launch vehicle's stability—that is, its resistance to deviation from its intended flight path—is related to its ratio of height to width; the higher the ratio, the greater the potential for problems with stability. Launcher designs usually attempt to maintain a height-to-width ratio that is no greater than 15:1; but, the five-segment stick will have a height-to-width ratio of 18:1. Thus, NASA expects that maintaining stability control of the five-segment stick will present a greater challenge than that posed by existing launch vehicles.
- **A large second stage powered by a modified engine.** Since the 1960s, the largest second stages have had dry weights (that is, weight that does not consist of fuel) of about 4 metric tons (mt) or less. By contrast, the second stage of the five-segment stick will weigh about 10 mt. (That additional weight is a contributor to the five-segment stick's relatively large height-to-width ratio.) The second stage will be powered by an updated version of the J-2S engine, which was developed at the end of the Apollo program.

Box 2.1. (Continued)

- **Structural integrity of the five-segment solid rocket booster.** The crew carrier's five-segment solid rocket booster must support the weight of both the second stage and the crew exploration vehicle during powered flight. NASA expects that the structural safety margins incorporated in the design of the four-segment sticks used for the space shuttle will suffice for the five-segment stick, even though the four-segment stick was designed for substantially different loads and stresses. However, the testing needed to verify that expectation remains to be done.

- **Organizing and executing the construction of a launch vehicle rated for manned space flight.** NASA last oversaw such an effort more than 20 years ago with the development and testing of the space shuttle.

Since NASA's publication of the Exploration Systems Architecture Study (ESAS), the crew carrier has evolved from a four-segment stick with a second stage powered by a space shuttle main engine (SSME) to a five-segment stick with a variant of the J-2S engine, which has less lift capability.

Evolution of the cargo carrier design has also begun. Recently, NASA announced that the SSME powering the cargo carrier recommended in the ESAS report will be replaced with the RS-68.[2] RS-68 engines offer lower cost and higher thrust; however, they also are less efficient (on a thrust-per-weight basis) than the SSMEs. That limitation could require the use of a powered second stage on the cargo carrier. (The Longfellow also features RS-68 engines in the first stage but adds a liquid-fueled second stage to overcome that limitation.) CBO lacks a detailed description of the evolving design of the cargo carrier and, therefore, has not considered that recent development among its alternatives.

[1] See, for instance, comments from NASA's manager of Constellation Systems Launch Vehicles Project at the Marshall Space Flight Center, cited by Frank Morring Jr. in "Tight Schedule; Crew Launch Vehicle Targets 2012 Human Mission; Challenges Abound," *Aviation Week & Space Technology*, April 24, 2006, p. 32.

[2] Frank Morring Jr. in "Bigger Prop Tank Was Key to NASA's RS-68 Decision," *Aerospace Daily & Defense Report*, May 22, 2006.

To execute the less ambitious schedule using super-heavy launchers, CBO estimates that costs through 2017 would range from $20 billion to $26 billion. CBO's estimates in this case indicate that using NASA's choice of launch vehicles would cost about $10 billion less through 2017 than the funding contained in NASA's budget projections.[46] (However, given the risks inherent in the designs of the crew and cargo launchers chosen by NASA, CBO's estimates could be low; see Box 2.1.) That savings comes from reducing the number of ISS support missions from 25 to 15 and delaying procurements for four lunar missions beyond 2017.

Appendix A On-Orbit Assembly, Launch Vehicle Reliability, and Overall Probability of Mission Success

On-Orbit Assembly

Relying on multiple launches to execute a manned lunar mission requires the capability to assemble various hardware subcomponents into a functional whole in low earth orbit (LEO). Such assembly requires that the subcomponents be designed to rendezvous in space and dock with one another. Those operations can be accomplished by humans (either remotely from the ground or locally in LEO) or autonomously. In constructing the International Space Station (ISS), the National Aeronautics and Space Administration (NASA) has gained experience assembling a large mass in orbit using a crew on-site. (The space shuttle was used to lift into LEO subcomponents of the ISS that were later assembled by the shuttle's crew.) During all of the Apollo lunar missions, the command and service module and the lunar module performed undocking and redocking maneuvers while in flight. Those operations were controlled locally by the crew in the command and lunar modules.

The Russians have demonstrated the ability to rendezvous and dock in space using remote control. In particular, since 1989 their Progress program has flown the Progress M and the Progress M1 on 55 flights to the former Mir space station and to the ISS with only one failure. (In 1997, the Progress M collided with the MIR and damaged it, although not catastrophically.) Incapable of carrying a crew, the Progress vehicles delivered cargo to either the MIR or the ISS.

The United States has not successfully demonstrated rendezvous and docking by remote control or autonomously. NASA launched a Demonstration of Autonomous Rendezvous Technology (DART) spacecraft in April 2005 to rendezvous with and conduct close-proximity maneuvering about a target spacecraft. However, the mission terminated early, and NASA reported that the DART spacecraft had experienced, among other problems, irregularities in its navigation system.[47] The Department of Defense is pursuing initiatives similar to DART. One is the Orbital Express program, which is designed to demonstrate on-orbit satellite servicing; it is scheduled for launch in late 2006. Another is the XSS-11 mission, which successfully rendezvoused with its launch vehicle in April 2005 and is scheduled to spend a year rendezvousing (but not docking) with other objects in orbit.

With no directly relevant experience, it is difficult to assign a probability of success to eventual assembly operations that might be conducted by the United States—either remotely or autonomously—in an effort to return humans to the moon. Using the demonstrated success of the Russian Progress program—0.982—as a benchmark, CBO performed a parametric analysis of the likelihood of a manned lunar mission's failing using probabilities of success for on-orbit assembly that ranged from 0.98 to 0.99. In that analysis, the crew-assisted docking that would follow the launch of the crew carrier spacecraft is assumed to be successful.

Launch Reliability

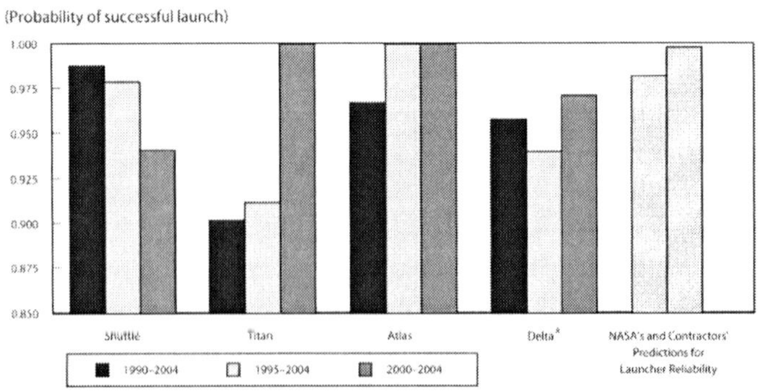

Source: Congressional Budget Office.

a. The Delta IV heavy-lift launcher, included in these averages, accomplished test objectives but failed to place its dummy payload into the intended orbit.

Figure A.1. Historical Launch Reliability

CBO's parametric analysis uses probabilities of successful launch ranging from 0.959 to 0.998. The lower bound of that range is the 15-year average of successful launches of the Atlas, Delta, and Titan expendable launchers, as well as the space shuttle (see Figure A.1). The upper bound is the average of the probabilities of successful launches projected by Boeing, Lockheed Martin, NASA, and ATK Thiokol for the launch vehicle options that each organization considered during 2005 prior to completion of NASA's Exploration Systems Architecture Study.

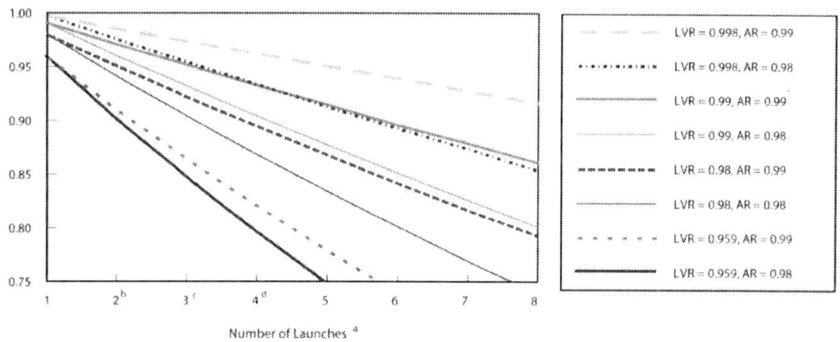

Source: Congressional Budget Office.

Notes: AR = assembly reliability; LVR = launch vehicle reliability.

From 1990 to 2004, the average overall LVR of the shuttle and the Atlas and Delta families is 0.959.

a. The number of launches is N-1 cargo launches plus 1 crew launch. Reliability for docking the crew exploration vehicle with its cargo is assumed to be 1.0.

b. Probability of mission success with the super heavy launchers.

c. Probability of mission success with the shuttle-derived Side-Mount, Longfellow, or Atlas V Heavy Plus cargo vehicles.

d. Probability of mission success with the Delta IV Heavy Plus cargo vehicle.

Figure A.2. Overall Probability of Mission Success

Overall Probability of Launch and Assembly Success

Not surprisingly, alternatives that require the least number of launches are least likely to experience a failure (see Figure A.2). The use of close-derivative launchers to execute a manned lunar mission would require three or four launches, with a probability of failure ranging from 3 percent to 21 percent.

The use of the super-heavy alternatives, which would require two launches, would have a probability of failure of 2 percent to 10 percent.

APPENDIX B HUMAN SAFETY ON LAUNCH ABORT

Human safety on launch abort is a primary consideration when distinguishing between the lift capacity of crew carriers and cargo carriers. As mentioned in Chapter 1, maximizing crew safety while maximizing launch vehicle lift capacity can be conflicting goals. No crew carrier considered in this Congressional Budget Office (CBO) report would fail to meet safety standards set by the National Aeronautics and Space Administration (NASA). However, trade-offs in design must occur to ensure crew safety, and those trade-offs affect the alternatives differently. Human safety during launch abort becomes an issue only if a mission has to be terminated because the launch has already failed to the degree that the crew must return to Earth safely before it reaches orbit.

Launch Trajectories and Launch Abort

The potential for death or injury during an aborted launch depends on the deceleration that the crew experiences as the spacecraft returns to Earth; that deceleration, in turn, depends on the launcher's trajectory on ascent. A launch vehicle that lifts its payload to a high altitude quickly is said to fly a steep trajectory; a launcher that ascends to a high altitude more gradually is said to fly a shallow or depressed trajectory. In general, the more depressed the trajectory that a booster flies, the more work it has to do to lift its payload into orbit. Therefore, a given launcher will be able to deliver less payload to orbit if it flies a more depressed trajectory. However, more depressed trajectories are safer for crewmembers, if they must abort, because they fall less steeply back into the atmosphere. Hence, trajectories that tend to maximize payload delivery tend to reduce crew safety.

Both the magnitude of the deceleration that the crew experiences—measured in multiples of the acceleration of gravity at the Earth's surface, or g's—and the duration of the acceleration are relevant in determining whether a return trajectory might harm or kill the crew of a spacecraft. On the basis of data collected by the Air Force during tests of aircraft ejection seats, NASA has developed a standard for acceleration and its duration along the "eyeballs in" direction that should not be equaled or exceeded during a launch abort in order to ensure the survival of the spacecraft's crew (see Figure B.1).[48]

A Comparison of the Alternatives

To be deemed safe, a launch vehicle must fly an ascent trajectory that, in the case of launch abort, would result in a return trajectory that satisfied NASA's safety standard. (As illustrated in Figure B.1, the vehicle would fly trajectories with curves of acceleration versus duration that lie below the curve labeled "NASA Standard.") Because the existing evolved expendable launch vehicles (EELV) lift their unmanned cargo payloads along relatively steep trajectories, the EELV-derived crew carriers considered by CBO generally reduce payload capability and adopt depressed trajectories to maintain a safe profile for manned flight. The space shuttle's abort trajectory has substantial margin in meeting NASA's safety standard, as would the five-segment, shuttle-derived crew carrier.

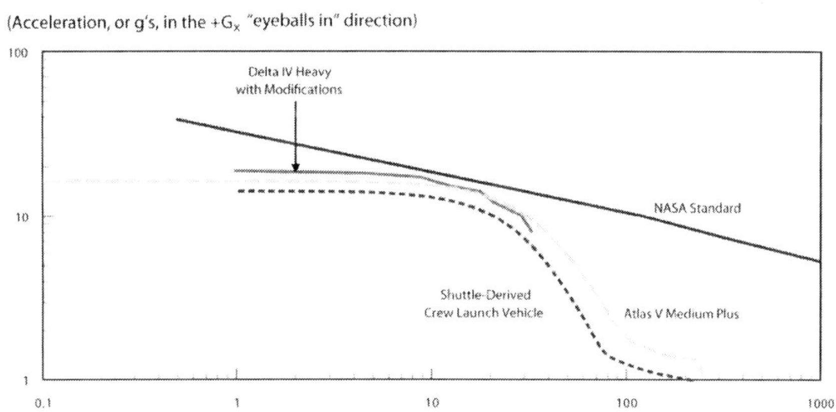

Sources: Congressional Budget Office based on contractors' estimates for launch vehicles. The NASA standard is from National Aeronautics and Space Administration, *Man-Systems Integration Standards*, NASA-STD-3000, vol. 8, *Crew Exploration Vehicle Launch Segment* (2005).

Note: Force is expressed in multiples of terrestrial surface gravity (g's). The direction of force is eyeballs in (with deceleration pressing on the astronauts' chests and eyes as they lie on their backs while descending toward Earth).

Figure B.1. Acceleration on Launch Abort for Various Launch Vehicles

Analogous calculations for the Saturn V launcher were performed for CBO by the staff of ATK Thiokol. Those calculations indicate that if a launch abort had occurred during the Apollo program, the Saturn V launcher's trajectory on ascent could have subjected the Apollo command module's crew to unsafe decelerations during its return to Earth (see Figure B.2).

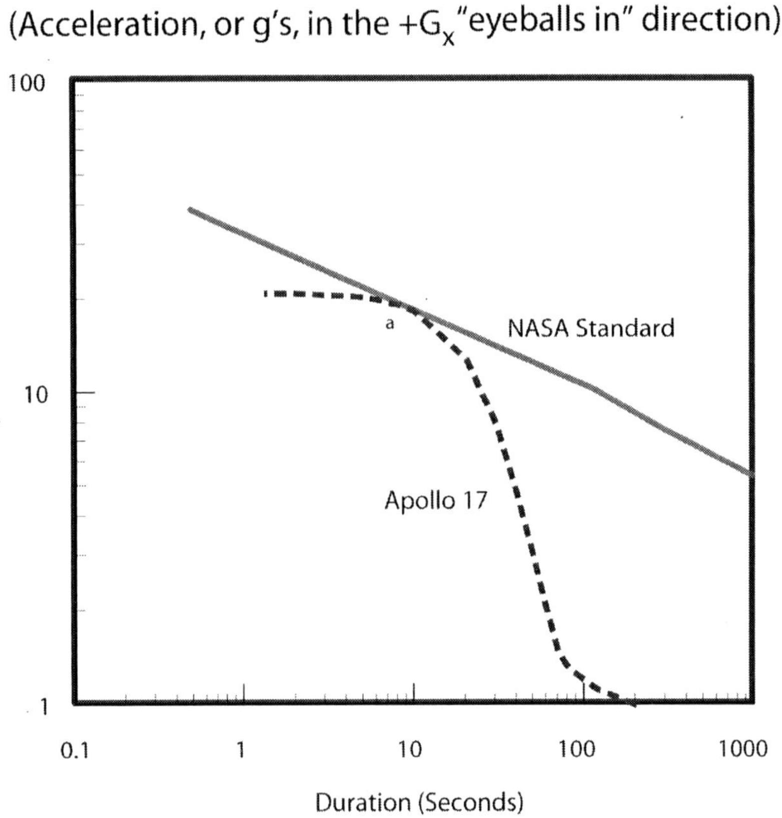

Sources: The NASA standard is from National Aeronautics and Space Administration, *Man-Systems Integration Standards*, NASA-STD-3000, vol. 8, *Crew Exploration Vehicle Launch Segment* (2005). The Apollo 17 calculations were done for the Congressional Budget Office by ATK Thiokol.

Note: Force is expressed in multiples of terrestrial surface gravity (g's). The direction of force is eyeballs in (with deceleration pressing on the astronauts' chests and eyes as they lie on their backs while descending toward Earth).

a. Apollo 17's maximum abort acceleration slightly exceeds the NASA standard.

Figure B.2. Acceleration on Launch Abort for Apollo 17

Appendix C Considerations That Affect Lunar Missions Executed Using Multiple Launches

Alunar mission that relied on multiple launches into low earth orbit (LEO) would be subject to constraints that would not affect missions conducted with a single Apollo-style launch. A primary constraint is the need to ensure that all the launches were executed in a short enough span of time to preclude substantial leakage of the cryogenic hydrogen fuel stored in orbit, which would be used to launch the assembled spacecraft from LEO to the moon. That constraint would make it necessary to account for potential launch delays and manage actual delays caused by problems related to weather, range safety issues, and launch vehicle performance. The potential need to launch other spacecraft from the same launch pad(s) used for launching manned lunar missions could cause scheduling constraints, which also must be considered.

Hydrogen Leaks

Hydrogen molecules are the lightest in existence and, therefore, are quite volatile. Consequently, anything containing hydrogen, even in the cold, near-vacuum conditions of space, leaks readily.[49] Assuming a predetermined mission length and launch schedule, launch vehicle designers plan for the excess hydrogen fuel needed to compensate for the leakage. (For example, about 34 percent of the mass of hydrogen fuel stored in LEO might leak from its on-orbit storage container during one year.)[50] However, once a payload is placed into LEO with a given amount of cryogenic hydrogen fuel on board, other launches must proceed according to the assumed schedule or too much hydrogen will boil away, leaving insufficient amounts of fuel to propel the assembled lunar mission out of LEO and to the moon.

Launch Cycle

A launch cycle generally consists of two phases: the time that is required at a launch complex to assemble and ready the launch vehicle and payload involved in a single launch, and the time required to both execute the launch and refurbish the complex for the next anticipated launch. Infrastructure at the shuttle launch complex and at the launch complex that serves the Atlas evolved expendable launch vehicle (EELV) can support one launch every 30 days. The launch cycle for the Delta EELV launch complex is 21 days. Therefore, as currently configured and if fully staffed, the shuttle and Atlas complexes could each support 12 launches annually, and the Delta complex, 18.

Launch Delays

Most delays, or at least their duration, can be anticipated and accommodated within the launch cycle. Approximately one-third of Air Force launches occur on time. Another third of launches occur with a delay that can be accommodated within the launch cycle. And roughly one-third of launches are delayed until later launch cycles.[51] About 50 percent of those launch delays are caused by weather, 42 percent are caused by malfunctioning hardware (in the launch vehicle, payload, or sensing systems prior to launch), and 8 percent are caused by range problems (such as the failure of a tracking camera or other sensor used to monitor the launch downrange after the vehicle's launch or prior to its reaching orbit).

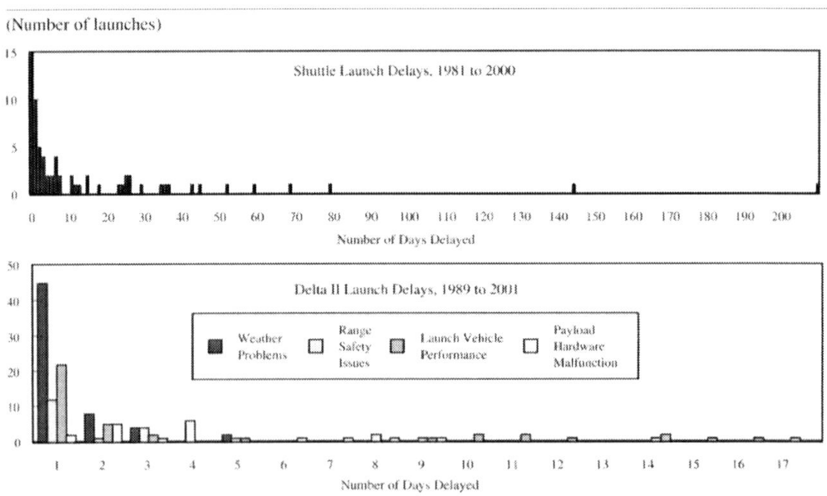

Source: Congressional Budget Office based on data from shuttle-mission-delay records and from D.P. Thunnisen, *Balancing Cost, Risk, and Performance Under Uncertainty in Preliminary Mission Design* (Reston, Va.: American Institute of Aeronautics and Astronautics, 2004).

Figure C.1. Historical Launch-Pad Delays for Selected Launch Systems

The typical delay is usually of short duration. Simpler launchers like the Delta II have experienced an average launch delay of only 2.4 days, whereas more complex launch systems can suffer longer delays (see Figure C.1). The shuttle's average delay has been 12 days.

To accommodate such delays, launch cycle missions contain built-in margins. For instance, for the Atlas EELV, a four-day margin is built into the schedule. In addition, a typical launch cycle will assume that work is conducted five days per week, with no work performed on weekends. Therefore, a 30-day launch cycle may have a built- in allowance for delays of up to about 12 days.

Experience indicates, however, that some very long launch delays can occur, particularly for the more complex launchers (see Figure C.1). A less complex launcher like the Delta II experiences launch delays that exceed 12 days about 4 percent of the time. The more complex space shuttle has experienced launch delays that last longer than 12 days about 20 percent of the time, but it also has experienced delays of longer than 30 days about 10 percent of the time.

Table C.1. Launch Cycles Needed to Ensure Mission Success

T	L		D	L + D
		p = 0.9		
0.99	4		3	7
0.99	2		2	4
0.999	4		4	8
0.999	2		3	5
		p = 0.85		
0.99	4		4	8
0.99	2		3	5
0.999	4		5	9
0.999	2		4	6

Source: Congressional Budget Office based on data provided by Boeing, Lockheed Martin, and ATK Thiokol.

Note: T = threshold of success (percent); L = required launches per mission; D = number of additional launch cycles planned for delays; L + D = number of launch cycles necessary for mission success to exceed T; p = probability of launch within a given launch cycle, which is assumed to be equal for all launches associated with a given mission.

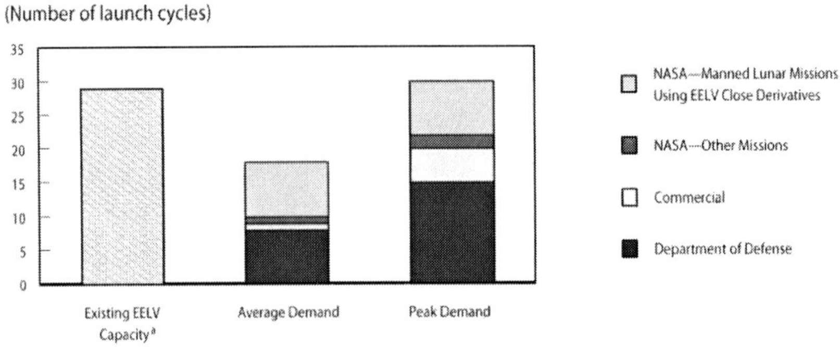

Source: Congressional Budget Office.

Notes: EELV = evolved expendable launch vehicle; NASA = National Aeronautics and Space Administration.

The "Average Demand" column displays the arithmetic average of the demand projected for the period spanning 2005 through 2020 for Department of Defense and commercial launches; the arithmetic average of the demand projected by NASA for launches of other than manned lunar missions for the period spanning 2005 to 2012; and CBO's estimate of the launches needed to execute manned lunar missions through 2020 using the pure Delta-derived alternative. The term "pure" means that both the crew carrier and the cargo carrier are derived from vehicles in the same family of launchers.

The "Peak Demand" column displays the maximum annual demand projected for each mission over the time periods considered.

a. A potential surge capacity of up to about 35 launch cycles is possible.

Figure C.2. EELV Annual Launch Capacity and Potential Demand

Launch Delays and Launch Cycles

Because there is some chance that complex launch vehicles will experience a long delay, the total number of launch cycles available to lift the components of a manned lunar mission can become a key consideration.

This is particularly true if the number of mission launches is large. In general, the number of launch cycles available for existing launchers would be sufficient to accommodate both manned lunar missions and other launches, providing the former occurred without appreciable delay. However, if delays became too long, then current launch capacity could be insufficient. The following calculation indicates the interplay between launch cycle capacity and launch delay.

Let:

L = the number of launches required for a manned lunar mission (*L* ranges between 2 and 4 for the alternative launchers considered by the Congressional Budget Office);

B = the boil-off constraint in launch cycles (*B* is the number of launch cycles within which hydrogen boil-off is not a constraint. For instance, if the launch cycle is one month, and boil-off becomes a problem after eight months, then *B* = 8);

D = the number of additional launch cycles put into a mission's launch schedule to accommodate the possibility of delays;

p = the probability that a given launch will occur within one launch cycle (for example, the data for the shuttle and Delta II suggest that launches of manned lunar missions might occur within the planned launch cycle about 90 percent of the time, or *p* = 0.9); and

T = the desired probability that all launches occur on schedule (for instance, *T* = 0.99 would reflect the desire that all *L* launches occur within *L* + *D* launch cycles at least 99 percent of the time).

Using that framework, the minimum number of added launch cycles, *D*, that must be built into a launch schedule to ensure the desired probability, *T*, of executing all launches on schedule can be computed as a function of the number of launches needed, *L*, and the probability that a single launch occurs within one launch cycle, *p*.

In equation form, the relationship between these variables can be expressed by the inequality:

$$T \leq p^{L} \sum_{i=0}^{D} (1-p)^{i} \binom{L+i-1}{i}$$

If it is desired that *T* be 0.99 or greater, the total number of launch cycles that must be incorporated (*L* + *D*) in constructing a launch schedule for a manned lunar mission would be about twice, or more, the number of launches needed (*L*). (See Table C.1.)

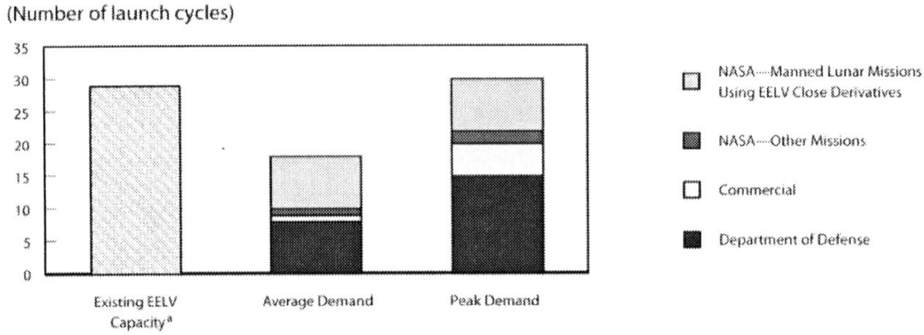

Source: Congressional Budget Office.

Notes: EELV = evolved expendable launch vehicle; NASA = National Aeronautics and Space Administration.

The "Average Demand" column displays the arithmetic average of the demand projected for the period spanning 2005 through 2020 for Department of Defense and commercial launches; the arithmetic average of the demand projected by NASA for launches of other than manned lunar missions for the period spanning 2005 to 2012; and CBO's estimate of the launches needed to execute manned lunar missions through 2020 using the pure Delta-derived alternative. The term "pure" means that both the crew carrier and the cargo carrier are derived from vehicles in the same family of launchers.

The "Peak Demand" column displays the maximum annual demand projected for each mission over the time periods considered.

a. A potential surge capacity of up to about 35 launch cycles is possible.

Figure C.2. EELV Annual Launch Capacity and Potential Demand

There are two additional constraints on *D*. One expresses the requirement that $L + D$ not exceed *B*, or $L + D <= B$. The results displayed in Table C.1 indicate that planning to launch enough hydrogen fuel to accommodate six months of leakage ($B = 6$) might be an insufficient margin if the close-derivative launchers ($L = 3$ or 4) considered by CBO were used to execute a manned lunar mission from the same launch pad. That margin would probably be sufficient, however, if most of the super-heavy launchers considered by CBO were used ($L = 2$, which excludes the Longfellow).

A second constraint on *D* is that consecutive launch cycles not conflict with other launch demand. For all super-heavy alternatives and the shuttle-derived close derivatives, the launch pads and other facilities used would be dedicated exclusively to launching lunar missions; therefore, conflict would not be an issue. For the EELV close-derivative launchers considered by CBO, however, conflict would be an issue because those launchers would use the same launch pads and facilities now used for launching Atlas and Delta EELVs. In particular, during years in which demand peaks for EELV launches that support other than manned lunar missions, existing capacity would be barely sufficient to accommodate the demand for those launches as well as the EELV launches needed to conduct manned lunar missions (see Figure C.2). In case a single launch vehicle proved unreliable, that near sufficiency could put the Air Force's desire for assured access at risk.[52]

APPENDIX D SCHEMATIC DEPICTIONS OF LAUNCH VEHICLE DERIVATIVES AND THEIR ANTECEDENTS

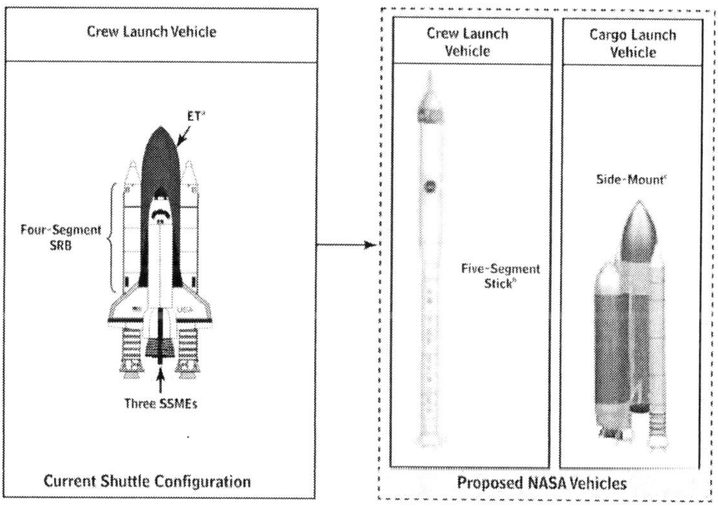

Source: Congressional Budget Office based on information provided by Boeing, Lockheed Martin, and ATK Thiokol.

Note: ET = external tank; SRB = solid rocket booster; SSMEs = space shuttle main engines.

a. The main propulsion system for the shuttle is composed of three "sticks": two four-segment SRBs attached to the ET (the third stick).

b. Modifications would include increasing the four-segment SRBs to five segments each, adding a new upper stage using a simplified version of the J-2 engine (J-2S) from Saturn V, and adding a new launch vehicle human monitoring system.

c. Modifications would include using the existing SRBs and ET, and developing a new, expendable payload carrier with three simplified SSMEs.

Figure D.1. Proposed Modifications to Generate the Shuttle Close Derivatives

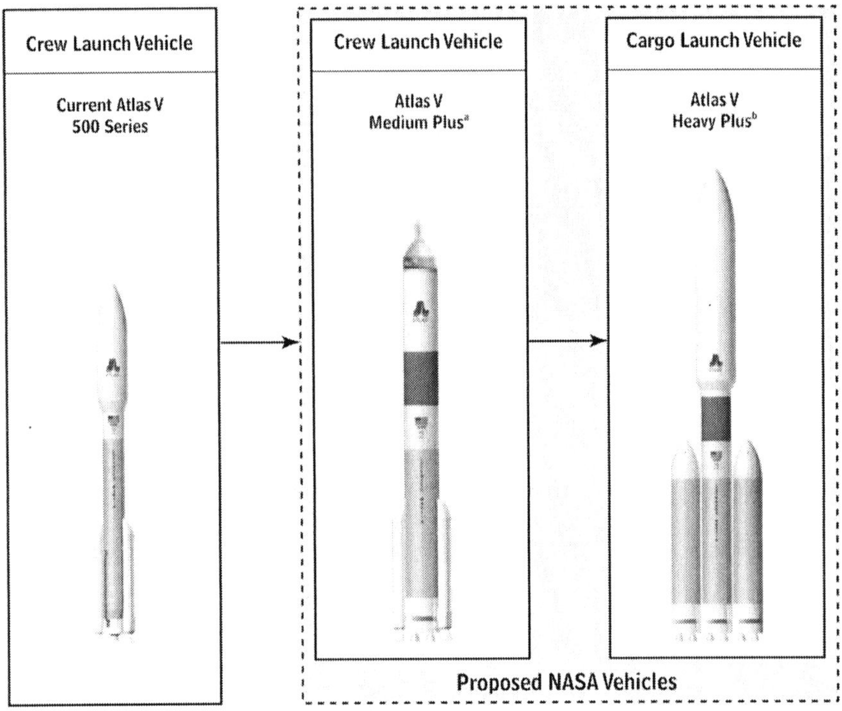

Source: Congressional Budget Office based on information provided by Lockheed Martin.

a. Modifications would include a new 5.4-meter first-stage booster tank and an upper-stage tank, two RD-180 engines for a booster, and four RL-10 engines for the upper stage. Reliability enhancements and a new launch vehicle human monitoring system would be added.

b. The Atlas V Heavy Plus would consist of a three-body version of the Atlas V Medium Plus, the same booster design, and the same upper- stage design. An expanded cargo shroud would be included.

Figure D.2. Proposed Modifications to Generate the Atlas V Close Derivatives

The figures in this appendix—which, within a given figure, are drawn to scale—depict the crew and cargo launch vehicles considered by the Congressional Budget Office under various alternatives. Their antecedent launch systems are also included.

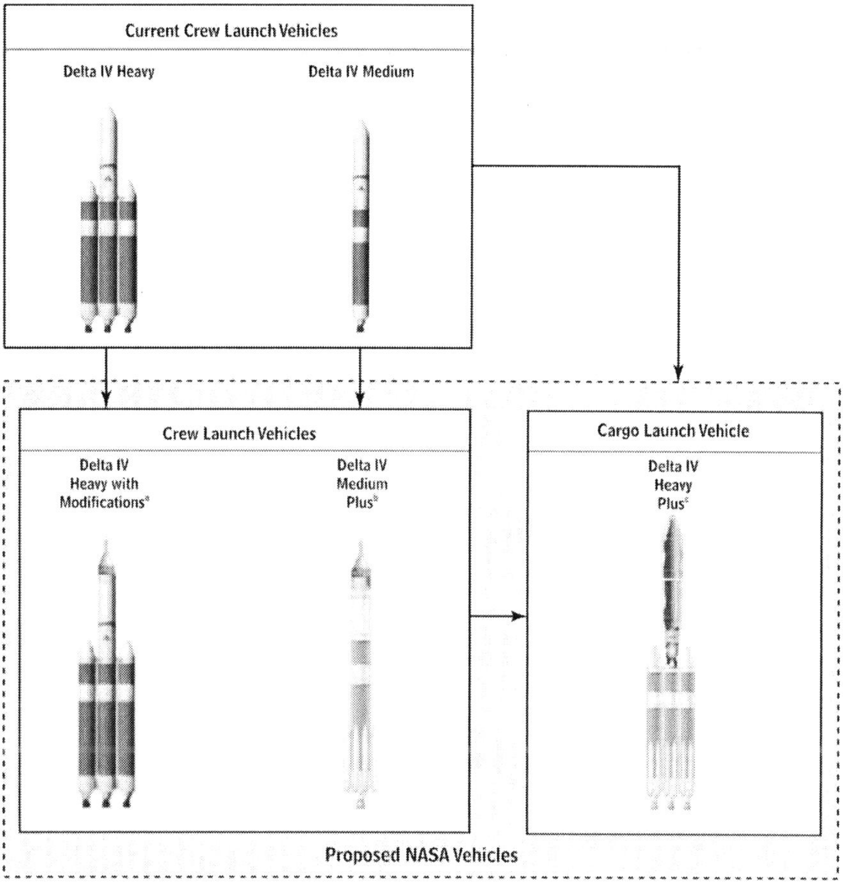

Source: Congressional Budget Office based on information provided by Boeing.

a. Modifications would include increased redundancy and new launch vehicle human monitoring systems, but the same cores and second stage would be used.

b. Modifications would include increased redundancy, new launch vehicle human monitoring systems, an enhanced first stage and six graphite epoxy motors (GEMs), and an enhanced second stage with three RL-10 engines.

c. The Delta IV Heavy Plus would include six GEMs, the same upper stage as the Delta IV Medium Plus, and a larger payload shroud.

Figure D.3. Proposed Modifications to Generate the Delta IV Close Derivatives

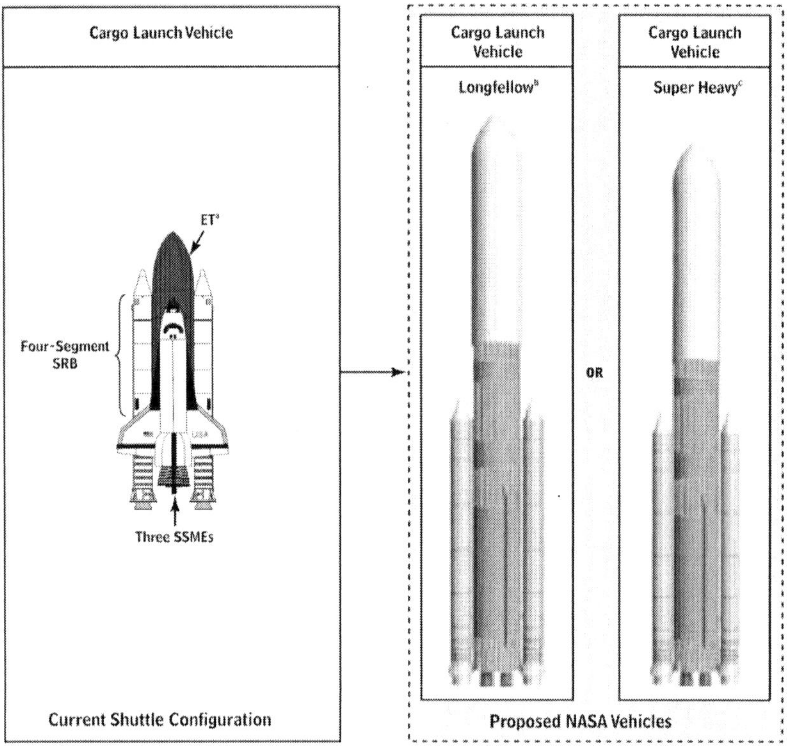

Source: Congressional Budget Office based on information provided by NASA and Boeing, Lockheed Martin, and ATK Thiokol.

Note: EDS = earth departure stage; ET = external tank; SRB = solid rocket booster; and SSMEs = space shuttle main engines.

a. The main propulsion system for the shuttle is composed of three "sticks": two four-segment SRBs attached to the ET (the third stick).

b. Modifications would include using five-segment SRBs, developing a new first stage from a modified and stretched ET, and four RS-68 engines. A new second stage using one J-2S engine would be developed along with a new expendable payload carrier mounted above the new second stage. With its second stage, the Longfellow is about 393 feet high and about 15 feet taller than the super heavy.

c. Modifications would include using five-segment SRBs, developing a new first stage from a modified and stretched ET, and five SSMEs. The rocket would have no second stage (circular orbit requires a first burn of the EDS). A new expendable payload carrier mounted above a modified ET would also be developed.

Figure D.4. Proposed Modifications to Generate the Shuttle-Derived Super-Heavy Launchers

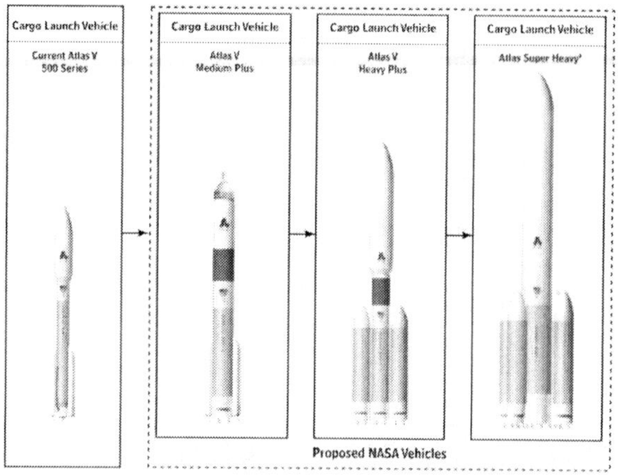

Source: Congressional Budget Office based on information provided by Lockheed Martin.

a. Modifications to the first stage would include a new 8.4-meter central core with five RD-180 engines and four liquid rocket boosters (LRBs) that are similar to the Atlas V Heavy Plus's LRBs. The second stage would include a Centaur second stage that is typical of the Atlas V Heavy Plus. A lengthened payload shroud would be added.

Figure D.5. Proposed Modifications to Generate the Atlas Super-Heavy Launchers

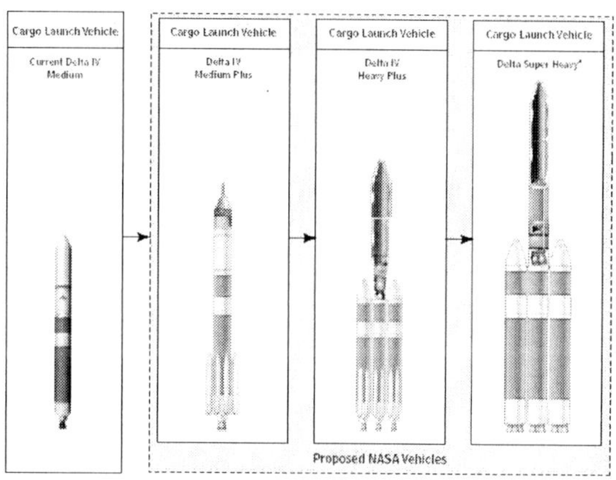

Source: Congressional Budget Office based on information provided by Boeing.

a. Modifications to the first stage would include three new 8-meter cores, each with four RS-68 engines. The second stage would consist of a new 8-meter stage with space shuttle main engines and a new, expanded 8-meter cargo shroud.

Figure D.6. Proposed Modifications to Generate the Delta Super-Heavy Launcher

Appendix E Cost Estimates for Alternative Launch Systems

In this study, the Congressional Budget Office (CBO) examined two time frames that would enable the National Aeronautics and Space Administration (NASA) to meet future space-launch needs designed to further both human exploration of the solar system and a return to the moon. Each schedule takes into account a Presidential directive issued in January 2004 that called for a return mission to the moon no later than 2020. Subsequent to that directive, NASA concluded that a lunar mission in 2018 was feasible and announced plans to launch two lunar missions a year from that point forward. In addition, NASA has stated publicly that the new launch system used for lunar missions would also be used to service the International Space Station (ISS). On the basis of information provided by NASA, CBO estimates that between 15 and 25 ISS support missions would be launched.

The first path considered by CBO would achieve the more ambitious goal of two lunar missions a year by 2018 and would provide 25 ISS missions over the 2012-2016 period. The second path assumes that the moon missions would not start until 2020 and that the number of ISS missions, starting in 2014, would total only 15. For both paths, CBO considered six program alternatives that would meet the specific mission goals. Each alternative would involve purchasing two types of launch vehicles—one for transporting the crew and supporting the ISS, and a second, larger vehicle (called the cargo carrier) for lifting the rest of the lunar payload into orbit.

This appendix details the basis for the cost estimates developed for each of the six program alternatives, assuming they would be implemented under the more ambitious schedule. (Tables, however, may include data for both schedules.) Under the more ambitious schedule, each alternative would call for the purchase of 29 crew launchers and from four to 12 cargo launchers, depending on the lift capability of each cargo launcher. Some of the cargo launchers represent new, larger systems that would be able to lift payloads weighing in excess of 100 metric tons (mt). As a result, fewer of them would be required. Others are closer derivatives of existing systems that on a single launch lift between 40 and 77 mt into orbit. A description of the configuration and costs of the existing launchers is provided for clarity and because those existing systems form the basis for the estimates of the larger launch systems.

CBO calculates that costs to execute the more ambitious schedule for each of the six alternatives would range from $26 billion to $38 billion. Those estimates include nonrecurring and recurring costs incurred over the 2006-2017 period. Nonrecurring costs include costs for modifying existing launch systems, designing and developing new launchers, and constructing or modifying launch pads for new launchers. Recurring costs include costs for purchasing launchers and launch services. All costs in this appendix are expressed in constant 2006 dollars.

Existing Launch Systems

Delta IV and Atlas V Launchers

The Delta IV is a two-stage launch vehicle built by the Boeing Corporation. Boeing produces five configurations of the launcher to place a mix of medium or heavy payloads into space. The first stage uses a booster core powered by an RS-68 engine, and the second stage is powered by an RL-10 engine. Both engines burn liquid hydrogen and liquid oxygen to generate thrust. The payload is encapsulated in a fairing for protection. The four Delta IV configurations that launch payloads weighing up to 13 mt into space use a single booster core that can be augmented with up to four solid rocket motors and a second stage that measures 4 or 5 meters in diameter. Those configurations are referred to as medium-lift launch vehicles.

The configuration that launches payloads weighing up to 25 mt uses three of the booster cores strapped together to provide greater lift and a 5-meter second stage. That configuration is called a heavy-lift launch vehicle.

The Atlas V, built by Lockheed Martin, is also a two-stage launcher. Lockheed Martin currently produces several configurations of the launcher that can place a mix of payloads weighing up to 20.5 mt into space. Those configurations are called Atlas V Medium (A5M) launch vehicles. Another configuration, designed but not being built, would be able to place heavier payloads into space. The A5M's first stage uses a booster core powered by an RD-180 engine, and the second stage is powered by one or two RL-10 engines. The RD-180 engine burns liquid oxygen and kerosene propellants to generate thrust, and the payload is encapsulated in a fairing for protection. All of the A5M configurations use a single booster core that can be augmented with up to five solid rocket motors and a second stage that measures slightly more than 3 meters in diameter. The configuration capable of launching heavier payloads would use three boosters strapped together to provide greater lift.

According to CBO's estimates, the cost to buy a medium-lift launch vehicle (defined in this appendix as a launcher capable of lifting fewer than about 20 mt) would be about $200 million: roughly $100 million to purchase the launcher and another $100 million to pay for launch services. The cost to buy a heavy-lift launch vehicle (capable of lifting about 25 mt) would total roughly $350 million: about $200 million to purchase the launcher and another $150 million to pay for launch services. The estimate for the medium-lift launch vehicle includes the cost of four solid rocket motors that would be strapped to the first-stage booster core. (For a breakdown of the estimates by cost category for each of the existing launchers, as well as estimates of the recurring costs for new launchers, see Table E.1.)

Information provided by the Air Force in budget justification materials to the Congress served as the basis for CBO's estimates. The Air Force calculates that the cost to buy 137 launch vehicles—125 medium and 12 heavy launchers—would total nearly $23 billion in 1995 dollars. After inflating that figure to 2006 dollars, CBO projects that the Air Force's estimate would increase to $28 billion. The Air Force has indicated that its estimated costs are almost evenly split between launchers and launch services, with about $15 billion paying for the launchers and $13 billion paying for launch services. Although the Air Force budget justification materials did not distinguish between the heavy launcher and medium launcher when considering costs, CBO was able to calculate separate estimates by assuming that each heavy launcher would cost about twice as much as a medium one. The factor of two assumes that the recurring cost for the booster core is about the same as the cost for the second stage. Therefore, a heavy launcher consisting of three booster cores and one second stage (four components) would cost twice as much as a medium launcher consisting of one booster core and one second stage (two components).

Similarly, CBO calculated the costs of placing a heavy launcher and a medium launcher into space. Based on information provided by the Air Force and Boeing, CBO developed a cost-estimating relationship that uses the weight of the launcher to determine costs for launch services. CBO concluded that launch service costs increase with launcher weight. Using that relationship, CBO estimates that the cost to place a heavy launcher (with a dry weight of about 85 mt) into space would be about 50 percent more than the cost of placing a medium launcher (weighing about 50 mt).

Space Shuttle

The space shuttle is made up of three main components—an orbiter, two solid rocket boosters (SRBs), and an external tank (ET). The orbiter serves as the crew's home in space and is powered by three space shuttle main engines (SSMEs) that burn liquid hydrogen and liquid oxygen to generate thrust. The cockpit is located in the forward fuselage, and the main engines are located in the aft fuselage. Mission payloads are carried in the midsection of the orbiter. The two SRBs are each powered by a solid rocket motor. The motors burn solid fuel mixed with oxygen and provide most of the thrust to launch the space shuttle. The external tank contains the propellants used by the three main engines on the orbiter.

On the basis of information that NASA provided to the Congress in 2002 in budget justification materials, the cost to launch a space shuttle is about $900 million, assuming a launch rate of five missions per year. Unit costs would be higher with lower launch rates. Some components of the space shuttle, namely the orbiter and the SRB canisters, are reusable, and NASA refurbishes them after each launch. Although less than the cost of purchasing new components, the cost to refurbish the orbiter and SRBs is significant.

New Launch Systems

As noted earlier, this CBO study examined six alternatives for purchasing new launch systems to support missions both to the moon and to the ISS. Four alternatives would involve buying launchers that either are derivatives of or use components of the Delta IV and Atlas V launchers; two alternatives would call for purchasing launchers that either are derivatives of or use components of the space shuttle. (For a summary of costs for the six alternatives assuming the more ambitious schedule, see Table E.2.) The costs vary from $26 billion to $38 billion. All the alternatives call for the purchase of 29 crew launchers—25 for ISS missions and four for lunar missions. Appropriations for those purchases would start in 2010. Depending on the alternative, either four, eight, or 12 cargo launchers would be purchased, starting in 2016.

The technical characteristics and costs for launchers that would be purchased for each alternative are provided below. In many cases, CBO employed cost-estimating relationships developed jointly by NASA and the Air Force that use the weight of a launcher's components to estimate the cost of those components. (The cost- estimating relationships are contained in the NASA-Air Force Cost Model, or NAFCOM.)

Table E.1. Comparison of Unit Recurring Costs for Selected Launchers

(Millions of constant 2006 dollars)				
Launcher	**Hardware**	**Launch Services**	**NASA Overhead**	**Total**
	Existing Launchers			
Delta IV Medium (D4M)	100	100	c	200
Atlas V Medium (A5M)	100	100	c	200
Delta IV Heavy (D4H)	200	150	c	350
Space Shuttle	b	b	b	900
	Possible NASA Crew Launchers			
Delta IV Medium Plus (D4M+)	150	150	100	400
Atlas V Medium Plus (A5M+)	150	150	100	400
Delta IV Heavy with Modifications (D4H with Mods)	200	150	100	450
Five-Segment Single Stick[a]	200	150	100	450
	Possible NASA Cargo Launchers			
Delta IV Heavy Plus (D4H+)	250	200	150	600
Atlas V Heavy Plus (A5H+)	250	200	150	600
Side-Mount	450	300	200	950
Longfellow	550	300	250	1,100
Delta Super Heavy	850	300	350	1,500
Atlas Super Heavy	600	300	300	1,200
Shuttle-Derived Super Heavy	700	300	300	1,300

Source: Congressional Budget Office.

Notes: NASA = National Aeronautics and Space Administration. The estimates of recurring costs assume a purchase rate of four or five units per year for existing launchers and new crew launchers and a rate of two a year for the new cargo launchers. The estimate of overhead costs for new launchers is calculated by multiplying the launcher and launch-service costs by 30 percent.

a. The main propulsion system for the shuttle is composed of three "sticks": two four-segment solid rocket boosters attached to the central external tank (the third stick).

b. Breakdowns of costs by category are not available.

c. These launchers do not incur NASA overhead costs because they are used by the Air Force to launch military payloads into space.

Alternative 1

According to CBO's estimates, 29 crew launchers and 12 cargo launchers would be purchased under this alternative (see Table E.2). The cost to develop and purchase those launchers would total about $28 billion over the 2006-2017 period, CBO estimates.

CBO assumed that the crew launcher would be a variant of the Delta IV Medium launcher, known as the Delta IV Medium Plus (D4M+). The D4M+ would use the same first-stage booster core currently used by the Delta family of medium launchers with six solid-fuel rocket motors. The D4M+ would increase the length of the second stage on the medium launchers by about 20 percent, adding two RL-10 engines. Because the new launch system would be used as a crew launcher, state-of-the-art features would be incorporated to aid human flight. Also, more safety features would be added to conform with NASA's standards for a manned launch.[53]

Allowing for about $3.5 billion in nonrecurring costs and roughly $12 billion in recurring costs, CBO estimates that the crew launcher would cost about $15 billion over all. The estimate of nonrecurring costs includes the following:

- about $1.4 billion for redesigning the current medium launcher to the D4M+ configuration, on the basis of information provided by the Delta IV contractor;
- about $1.1 billion for three test flights, using the recurring per-unit cost estimates that are discussed later in this appendix but increasing those estimates by 30 percent (a factor of 1.3) consistent with NAFCOM methods for estimating the cost of test-flight assets;
- about $0.3 billion for modifying launch pads, based on information provided by the contractor on the nature of the needed modifications;
- about $0.8 billion for the costs of program management and systems engineering—overhead—associated with NASA launch activities, assuming a 30 percent burden rate.[54] (For all estimates, CBO assumes that costs would be increased by 30 percent to account for NASA's overhead.)

The estimate of recurring costs of about $12 billion includes the cost of buying 29 D4M+ launchers, at an average cost of about $400 million each. As shown in Table E.1, the estimate of per-unit recurring costs includes about $150 million in launch vehicle costs, $150 million in launch service costs, and $100 million in overhead.

CBO assumed that the cargo launcher would be a variant of the Delta IV Heavy launcher, known as the Delta IV Heavy Plus (D4H+). The D4H+ would use the same first-stage booster cores and second stage currently featured on the Delta IV Heavy launcher but would add two solid-fuel rocket motors to each of the three booster cores and add two RL-10 engines to the second stage. Those changes would allow the new launch system to be used as a cargo launcher with a payload lift of about 40 mt.

According to CBO's estimates, the cargo launcher would cost about $12 billion—about $5 billion in nonrecurring costs and about $7 billion in recurring costs (see Table E.2). The estimated nonrecurring costs include the following:

- about $1.1 billion for redesigning the current heavy-lift launcher to the D4H+ configuration, on the basis of information provided by the Delta IV contractor;
- nearly $1.6 billion for conducting three test flights;
- about $0.3 billion for modifying the launch pads, on the basis of information provided by the contractor on the nature of the needed modifications;
- about $1 billion for "mothballing" many NASA launch facilities (as a result of not using Pad 39, now used by the space shuttle), on the basis of information provided by NASA; and
- about $1.2 billion for NASA overhead.

The estimate of recurring costs includes the cost of buying 12 D4H+ launchers, at an average cost of about $600 million each (see Table E.1 for detailed information on the breakout of recurring costs).

Alternative 2

Under this alternative, 29 crew launchers and eight cargo launchers would be purchased, CBO estimates (see Table E-2). The cost to develop and purchase those launchers would total about $26 billion over the 2006- 2017 period, CBO estimates.

CBO assumed that the crew launcher would be a variant of the Atlas V Medium launcher, known as the Atlas V Medium Plus (A5M+). The A5M+ would feature a redesign of the first-stage booster core and second stage currently used on the Atlas family of medium launchers. The width of those stages would increase from just over 4 meters to slightly more than 5 meters, but the length of the booster core would decrease by about one-third. The first stage would be powered by two RD-1 80 engines, and the second stage would be powered by four RL-10 engines. Because the new launch system would be used as a crew launcher, state-of-the-art features would be added to aid in human flight, and more safety features would be added to conform with NASA's standards for manned launch.

Table E.2. Summary of Costs for Space Exploration Alternatives from 2006 to 2017

(Billions of constant 2006 dollars)

Alternative	Set of Launchers	Purchase Quantity	Costs: One-Time	Costs: Recurring	Costs: Total
More Ambitious Schedule					
1	Crew: Delta IV Medium Plus (D4M+)	29	3.3	12.0	15.5
	Cargo: Delta IV Heavy Plus (D4H+)	12	5.2	7.2	12.4
	Total		**8.7**	**19.2**	**27.9**
2	Crew: Atlas V Medium Plus (A5M+)	29	5.3	12.0	17.4
	Cargo: Atlas V Heavy Plus (A5H+)	8	4.0	4.8	8.8
	Total		**9.3**	**16.8**	**26.1**
3	Crew: Five-Segment Stick[a]	29	4.8	13.5	18.3
	Cargo: Side-Mount	8	4.7	7.6	12.3
	Total		**9.5**	**21.1**	**30.6**
4	Crew: Delta IV Medium Plus (D4M+)	29	3.5	12.6	15.5
	Cargo: Delta Super Heavy	4	16.7	6.0	22.7
	Total		**20.2**	**18.0**	**38.2**
5	Crew: Atlas V Medium Plus (A5M+)	29	5.3	12.0	17.3
	Cargo: Atlas Super Heavy	4	9.0	4.8	13.8
	Total		**14.3**	**16.8**	**31.1**
6	Crew: Five-Segment Stick[a]	29	4.8	13.5	18.3
	Cargo: Shuttle-Derived Super Heavy	4	8.9	5.2	14.1
	Total		**13.7**	**18.7**	**32.4**

Table E.2. (Continued)

(Billions of constant 2006 dollars)					
			Costs		
Alternative	**Set of Launchers**	**Purchase Quantity**	**One-Time**	**Recurring**	**Total**
	Less Ambitious Schedule				
1	Crew: D4M+	15	3.5	6.0	9.5
	Cargo: D4H+	0	5.2	0.0	5.2
	Total		**8.7**	**6.0**	**14.7**
2	Crew: A5M+	15	5.3	6.0	11.3
	Cargo: A5H+	0	4.0	0.0	4.0
	Total		**9.3**	**6.0**	**15.3**
3	Crew: Five-Segment Stick	15	4.8	6.8	11.6
	Cargo: Side-Mount	0	4.7	0	4.7
	Total		**9.5**	**6.8**	**16.3**
4	Crew: D4M+	15	3.5	6.0	9.5
	Cargo: Delta Super Heavy	0	16.7	0.0	16.7
	Total		**20.2**	**6.0**	**26.2**
5	Crew: A5M+	15	5.3	6.0	11.3
	Cargo: Atlas Super Heavy	0	9.0	0.0	9.0
	Total		**14.3**	**6.0**	**20.3**
6	Crew: Five-Segment Stick	15	4.8	6.8	11.6
	Cargo: Shuttle-Derived Super Heavy	0	8.9	0.0	8.9
	Total		**13.7**	**6.8**	**20.5**

Source: Congressional Budget Office.

Note: The more ambitious schedule would involve purchasing 25 crew launchers to support the International Space Station (ISS) starting in 2010 and purchasing cargo launchers to support two lunar missions a year starting in 2016. The less ambitious schedule would involve buying only 15 crew launchers for the ISS missions starting in 2012 and would delay the start of purchases for the lunar missions to beyond 2017.

a. The main propulsion system for the shuttle is composed of three "sticks": two four-segment solid rocket boosters attached to the central external tank (the third stick).

According to CBO's estimates, the crew launcher would cost about $17 billion—roughly $5 billion in nonrecurring costs and about $12 billion in recurring costs (see Table E.2). The estimate of nonrecurring costs includes the following:

- about $2.2 billion for redesigning the current medium launcher to the A5M+ configuration, based on information from the Atlas V contractor;
- about $1.1 billion for conducting three test flights;
- about $0.3 billion for modifying launch pads;
- about $0.5 billion for equipping U.S. companies to manufacture all of the components of the Russian RD-1 80 engine; and
- about $1.2 billion for NASA overhead.

The estimate of recurring costs (about $12 billion) includes the cost of buying 29 A5M+ launchers, at an average price of about $400 million each, the same as the overall unit cost of the D4M+.

CBO assumed that the cargo launcher would be a variant of the Atlas V Medium launcher, known as the Atlas V Heavy Plus (A5H+). The A5H+ would use the same first- stage booster cores and second stage as the A5M+ crew launcher but would strap together three boosters in a manner similar to that used for the Delta IV Heavy launcher. Also, two RL-10 engines would be added to the second stage, bringing the total number of engines on the second stage to six. Those changes would allow the new launch system to be used as a cargo launcher with a payload lift of about 74 mt.

CBO estimates that the cargo launcher would cost about $9 billion—about $4 billion in nonrecurring costs and nearly $5 billion in recurring costs. The estimate of nonrecurring costs includes the following:

- about $0.2 billion to transition the A5M+ design to the A5H+ configuration, on the basis of information from the contractor regarding the needed redesign of the A5H+;
- about $1.6 billion for conducting three test flights;
- about $0.3 billion for modifying launch pads, on the basis of information provided by the contractor on the nature of the needed modifications;
- about $1 billion for deactivating NASA launch facilities; and
- about $0.9 billion for NASA overhead.

The estimate of recurring costs (about $5 billion) includes the cost of buying eight cargo launchers, at an average cost of about $600 million for each A5H+ (see Table E.1).

Alternative 3

CBO estimates that 29 crew launchers and eight cargo launchers would be purchased under this alternative. The cost to develop and purchase those launchers would total about $31 billion over the 2006-2017 period, CBO estimates.

CBO assumed that the crew launcher would use a variant of the space shuttle's solid rocket booster, known as the five-segment single stick. The five-segment single stick would be a two-stage launch vehicle. The first stage would be comparable to the SRBs used on the space shuttle, except it would add an extra segment of solid fuel to increase lift capability. A second stage would be added to further increase lift capability. The second stage would be powered by a variant of the J-2S engine.

According to CBO's estimates, the crew launcher would cost about $18 billion—nearly $5 billion in nonrecurring costs and over $13 billion in recurring costs (see Table E.2). The estimate of nonrecurring costs includes the following:

- about $0.8 billion for redesigning the first-stage solid rocket boosters, based on NAFCOM estimates;
- about $2 billion for designing a second stage and the associated shroud, based on information from NAFCOM;
- about $1.3 billion for conducting three test flights; and
- about $1.1 billion for NASA overhead.

The estimate of recurring costs (more than $13 billion) includes the cost of buying 29 five-segment, single-stick launchers, at an average price of about $450 million each.

CBO assumed that the cargo launcher would be a variant of the space shuttle, known as the Side-Mount. The Side-Mount would consist of three main components—two SRBs, an external tank, and a payload carrier. This variant would be most similar to the space shuttle, minus the orbiter and with a payload carrier replacing it. The payload carrier would be powered by three space shuttle engines that burn liquid hydrogen and liquid oxygen to generate thrust. The two SRBs would be identical to those used on the space shuttle, and each would be powered by a solid rocket motor. The external tank would contain the propellants used by the three main engines on the payload carrier. The payload carrier would be positioned on the side of the external tank.

CBO estimates that the cargo launcher would cost over $12 billion—nearly $5 billion in nonrecurring costs and nearly $8 billion in recurring costs. The estimate of nonrecurring costs includes the following:

- about $0.1 billion for redesigning the solid rocket boosters and external tank to allow them to interface with the payload carrier, on the basis of NAFCOM estimates;
- about $0.9 million to design the payload carrier, on the basis of NAFCOM estimates;
- about $2.6 billion for conducting three test flights; and
- about $1.1 billion for NASA overhead.

The overall cost of each Side-Mount launcher would total about $950 million, CBO estimates—about the same as the cost for each space shuttle. The cost for buying eight of the launchers would total nearly $8 billion, CBO estimates.

Alternative 4

According to CBO's estimates, 29 crew launchers and four cargo launchers would be purchased under this alternative (see Table E.2). The cost to develop and purchase those launchers would total about $38 billion over the 2006-2017 period, CBO projects.

CBO assumed that the crew launcher would be the D4M+, the same launcher that would be used under Alternative 1. As mentioned earlier, CBO estimates that the nonrecurring costs would total about $3.5 billion and the recurring cost for buying 29 D4M+ launchers would total about $12 billion.

CBO assumed that the cargo launcher would use components of the Delta IV Heavy launcher but would represent a new design with much greater lift capability. The D4 super heavy would feature redesigned versions of the booster cores and second stage currently used on the Delta heavy launcher, allowing the super heavy to be used as a cargo carrier with a payload lift of nearly 150 metric tons—almost six times the lift capacity of the Delta IV Heavy launcher. The length of the boosters would increase by 50 percent and the width would increase by 60 percent, resulting in a mass increase of about 150 percent. Like the Delta IV Heavy launcher, each D4 super heavy would use three booster cores strapped together, but each booster would be powered by four RS-68 engines, an increase of three engines per core over the current Delta IV cores. The length of the second stage would increase by 100 percent, and the width would increase by 60 percent, resulting in a mass increase of about 220 percent. The second stage would be powered by one SSME that would be modified from its current configuration on the space shuttle.

The cargo launcher would cost about $23 billion—about $17 billion in nonrecurring costs and about $6 billion in recurring costs, CBO estimates (see Table E.2). The estimate of nonrecurring costs includes the following:

- about $3.6 billion for redesigning the Delta IV Heavy launcher to the D4 super-heavy configuration, on the basis of information from the Delta IV contractor;
- about $4.3 billion for conducting three test flights;
- about $5 billion for modifying launch pads, on the basis of information provided by the contractor on the nature of the needed modifications to Pad 39; and
- about $3.9 billion for overhead.

The estimate of recurring costs of about $6 billion includes the cost of buying four of the cargo launchers, an average cost of about $1.5 billion for each launcher (see Table E.1 for detailed information on the breakout of the recurring costs).

Alternative 5

CBO assumed that 29 crew launchers and four cargo launchers would be purchased under this alternative. The cost to develop and purchase the launchers, according to CBO's estimates, would total about $31 billion over the 2006-2017 period.

CBO further assumed that the crew launcher would be a variant of the Atlas V Medium launcher, known as the A5M+—the same crew launcher considered under Alternative 2. As stated earlier, CBO estimates that the nonrecurring costs would total over $5 billion and that the recurring costs for buying 29 of those launchers would total about $12 billion.

Under this scenario, CBO assumed that the cargo launcher would use components of the A5H+—the cargo launcher considered under Alternative 2—but, known as the A5 super heavy, it would have much greater lift capability. The A5 super heavy would use the same type of boosters as those featured on the A5H+ but would strap four of the boosters into a cluster. Each booster would be powered by two RD-180 engines. In the center of that cluster would be a fifth booster, slightly longer and wider than the other four boosters and powered by five RD-1 80 engines. Thus, the first stage would be powered by a total of 13 RD-180 engines. (The second stage of the A5 super heavy would correspond to that of the A5H+.) Those changes would allow the vehicle to be used as a cargo carrier with a payload lift of about 134 mt.

CBO estimates that the cargo launcher would cost nearly $14 billion—about $9 billion in nonrecurring costs and nearly $5 billion in recurring costs. The estimate of nonrecurring costs includes the following:

- about $1.4 billion to transition the A5H+ design to the A5 super-heavy configuration, based on information provided by the contractor about the design of the A5 super heavy;
- about $3.3 billion for conducting three test flights;
- about $1.2 billion for modifying the launch pads, based on information provided by the contractor and NASA analysts;
- about $1 billion for mothballing NASA launch facilities; and
- about $2.1 billion for NASA overhead.

The estimate of recurring costs of over $5 billion includes the cost of buying four cargo launches at an average of about $1.2 billion each—about twice the per-unit cost of an A5H+ launch.

Alternative 6

As with all the alternatives considered under the more ambitious schedule, CBO estimates that 29 crew launchers would be purchased under this scenario; four cargo launchers would be required (see Table E.2). The cost to develop and purchase those 33 launch vehicles would total about $32 billion over the 2006-2017 period, according to CBO's projections.

CBO assumed that the crew launcher would be the five-segment single stick, the same vehicle considered under Alternative 3. As discussed earlier, CBO estimates that the cost of this launcher would total about $18 billion—about $5 billion in nonrecurring costs and over $13 billion in recurring costs.

CBO assumed that the cargo launcher would use components of the space shuttle but have much greater lift capability. The Shuttle-derived super heavy would be a one-stage launch vehicle consisting of three main components—two SRBs, an external tank, and a payload carrier. The two rocket boosters and the external tank would be strapped together to form the first stage. The rocket boosters would be identical to the booster used on the five-segment single-stick launcher. The external tank would be comparable to the tank used on the space shuttle and it would be powered by five SSMEs. Because those engines burn liquid hydrogen and liquid oxygen to generate thrust, placing five of them on the cargo launcher would provide lift capability of about 125 mt. The payload carrier would be positioned on top of the external tank.

The cargo launcher would cost about $14 billion, consisting of about $9 billion in nonrecurring costs and about $5 billion in recurring costs, according to CBO's estimates (see Table E.2). The projection of nonrecurring costs includes the following:

- about $1.1 billion for designing the external tank, on the basis of NAFCOM estimates;
- about $0.5 billion for designing the payload carrier, on the basis of NAFCOM estimates;
- about $3.5 billion for conducting three test flights;
- about $1.5 billion for modifying launch pads, on the basis of information provided by contractors about the nature of the needed modifications; and
- about $2 billion for NASA overhead.

The estimate of recurring costs of about $5 billion includes the cost of buying four cargo launchers, at an average of about $1.3 billion apiece (see Table E.1).

End Notes

[1] A metric ton is 1,000 kilograms, or about 1.1 short tons; a short ton is 2,000 pounds.

[2] Low earth orbit is defined as one of a group of circular orbits that lies between the appreciable atmosphere (which is about 120 nautical miles, or nm, high) and the Van Allen radiation belt (which is about 1,000 nm high) and has an inclination to the equator of less than 60 degrees. A variety of other orbit classes are recognized. Those with inclinations of greater than 60 degrees are referred to as polar orbits. Geosynchronous orbits (GSOs), which are also known as geostationary orbits, are about 22,000 nm high and allow satellites to remain over a single point above the Earth. Another class of orbits is the set of elliptical geosynchronous transfer orbits (GTOs) that are often used for transitions between LEOs and GSOs.

[3] CBO characterizes as close derivatives those new launch vehicles that are substantially similar to existing systems. The essentially new and much more capable launchers are termed "super-heavy" launchers.

[4] The evolved expendable launch vehicles are the Air Force's current fleet of launchers designed to take heavier payloads (weighing between 5 mt and 25 mt) into orbit.

[5] The weight that a launch vehicle can lift into orbit varies depending on the orbit and launch location. The orbit referenced here is circular, 220 nautical miles high, and inclined 28.6

degrees to the equator with launch from the Kennedy Space Center. For this particular orbit, the shuttle's performance is taken from S.J. Isakowitz and others, *International Reference Guide to Space Launch Systems*, 4th ed. (Reston, Va.: American Institute of Aeronautics and Astronautics, 2004), p. 434.

[6] See Office of Management and Budget, *Budget of the United States Government, Fiscal Year 2007* (January 2006).

[7] Although recently, NASA has announced that, instead of the SSMEs, it will be using RS-68 engines like those currently used on the Delta IV first stage.

[8] The term "pure" means that both the crew carrier and the cargo carrier are derived from vehicles in the same (Atlas-, Delta-, or shuttle-derived) family of launchers.

[9] In May 2006, NASA announced that its design for the cargo carrier had evolved and now featured RS-68 engines in an expanded first stage rather than SSMEs. Using that new design for the cargo carrier, CBO's preliminary estimates of the total programmatic costs for the more aggressive schedule decreased to $30 billion and, on the less aggressive schedule, to $19 billion.

[10] Typical unmanned payloads consist of satellites designed to track weather or facilitate communications, conduct scientific research, or perform surveillance.

[11] These projections are based on reports published by the Federal Aviation Administration and studies by the Futron Corporation, a company that the FAA and NASA use to project launch capacity and demand. See Federal Aviation Administration, Office of Commercial Space Transportation (AST) and the Commercial Space Transportation Advisory Committee, *2005 Commercial Space Transportation Forecasts* (May 2005); FAA and AST, *Commercial Space Transportation Quarterly Launch Report, 4th Quarter, 2004*; and Futron Corporation, *NASA ASCENT Study Final Report* (January 31, 2003). On April 11, 2005, Futron also provided data on projected launch vehicle rates directly to CBO. The projections displayed in this report include both capacity and demand for launching any payload expected to achieve earth orbit and exclude suborbital payloads (that is, payloads that do not achieve earth orbit). The capacity for space shuttle launches is assumed to be 10 annually. The FAA and Futron have published a May 2006 forecast for commercial launch demand that increases the potential number of such launches by seven in 2007 relative to the 2005 forecast. Beyond 2007, the 2005 and 2006 commercial launch forecasts differ very little.

[12] Geosynchronous orbits, also known as geostationary orbits, are about 22,000 nautical miles (nm) high and allow satellites to hover over points on the equator. Other classes of orbits are also recognized. A low earth orbit is defined as one of a group of circular orbits that lies between the appreciable atmosphere (which is about 120 nm high) and the Van Allen radiation belt (which is about 1,000 nm high) and has an inclination to the equator of less than 60 degrees. Orbits with inclinations greater than 60 degrees are referred to as polar orbits. Also recognized are elliptical geosynchronous transfer orbits that are often used for transitions between LEOs and GSOs.

[13] General T.S. Moorman (Ret.) and others, *Enabling Assured Space Access Study*, Booz Allen Hamilton (January 15, 2005), p. 17.

[14] The Delta IV Heavy has flown only once and failed to place its payload into the correct orbit. Versions of the Angara rocket under development by the Khrunichev State Research and Production Space Center in Russia are being designed to lift up to 23 mt. A version of the Russian Proton launch vehicle can lift 20 to 21 mt.

[15] The Single Economic Space consists of Russia, the Ukraine, Kazakhstan, and Belarus, which announced in the third quarter of 2004 that they intended to merge their space enterprises.

[16] Futron has completed sensitivity analyses on assumptions that underlie these observations, but with the possible exception of a "breakout" of space tourism, the excess in space-launch capability is expected to continue.

[17] Projections of U.S. launch capacity exclude the Sea Launch system operated by Boeing in partnership with RSC-Energia (based in Moscow), Kvaerner ASA (based in Oslo), and

SDO Yuzhnoye/PO Yuzhmash (based in Dnepropetrovsk, Ukraine). Those projections also exclude the capacity provided using Proton and Angara launch vehicles at the Bakonur Cosmodrome in Kazakhstan of International Launch Services, a joint venture between Lockheed Martin and the Khrunichev State Research and Production Space Center in Russia.

[18] The payload of 140 metric tons is derived from weight data provided in Richard W. Orloff, *Apollo by the Numbers: A Statistical Reference*, NASA SP-2000-4029 (National Aeronautics and Space Administration, updated September 27, 2005), available at http://history.nasa.gov/SP-4029/SP-4029.htm. In that reference, 140 mt is the weight of the Apollo 17 command-and-service modules, the lunar module, the spacecraft/lunar module adapter, the instrument unit, and the S-IVB stage (the third stage of the Saturn V), including the fuel remaining in that stage needed to propel the Apollo command-and-service modules and lunar module from low earth orbit to the moon.

[19] That estimate was made by NASA during the summer of 2005. As NASA and its contractors continue to work on the designs for the hardware that will be used to conduct manned lunar missions, such weight estimates are likely to change.

[20] The use of noncryogenic propellants has also been proposed as a way to eliminate the problem of fuel leaking while it is stored in LEO. However, cryogenic propellants such as liquid hydrogen burned with liquid oxygen are the most efficient in terms of thrust produced per pound of fuel burned per unit of time. Although the use of noncryogenic propellants can reduce the weight of fuel tanks and engines, more of such propellants is needed to launch a given payload from LEO to the moon than is needed using cryogenic fuels. The latter effect usually dominates, implying that the use of noncryogenic propellants would require a heavier payload to be lifted from the Earth into LEO. For example, CBO estimates that the use of hydrazine and nitrogen tetroxide rather than liquid hydrogen and oxygen as the propellant for launching a mission to the moon from LEO would about double the weight of the propellant that must be lifted into LEO. The additional weight of cryogenic fuel that would need to be lifted to provide sufficient margin to accommodate leakage is less if the multiple launches needed to execute a mission can be achieved within approximately six-month periods of time. For instance, with passive cooling techniques, it is possible to limit leakage to under 4 percent per month, or less than 30 percent in a six-month period, according to a 1982 study conducted for NASA. Modern insulation further limits losses. See National Aeronautics and Space Administration, Contractor Report 3536, *Future Orbital Transfer Vehicle Technology Study*, vol. 2 (1982).

[21] That choice was first presented with NASA's release of the draft results of its Exploration Systems Architecture Study (ESAS) in September 2005 and finalized in the ESAS *Final Report* in November 2005.

[22] Arguments favoring the separation of crew and payload are implicit in the recommendations contained in "Long-Term: Future Directions for the U.S. in Space," Chapter 9.3 in *Columbia Accident Investigation Board Report*, vol. 1 (August 26, 2003). As discussed in Appendix B of this report, maximizing crew safety while maximizing launch vehicle lift capacity can be conflicting goals.

[23] See National Aeronautics and Space Administration, *Launch Vehicle Capability Trade Study* (NASA briefing, 2003), which contains a comprehensive list of those nine studies. The lower-weight payloads correspond to the use of propulsion systems such as nuclear thermal rockets. Use of chemical propulsion of the same type that NASA now plans to use for the VSE lunar missions would require lifting payloads into LEO at the upper end of the range.

[24] Assuming that the propulsion systems and design constraints are similar, CBO estimates that launching a payload three times larger than the maximum capacity of the Saturn V would require a launch vehicle about 1.4 times taller and wider than the Saturn V. Even if all engineering issues were surmountable, the infrastructure to build, transport, and launch such a large vehicle does not exist. The costs for those new facilities, ignoring the costs of the large vehicles themselves, would be substantial.

[25] At the inception of the EELV program, the plan was to develop both Delta IV Heavy and Atlas V Heavy launchers capable of lifting about 25 mt into low earth orbit. Subsequently, the Department of Defense and Lockheed Martin agreed to forgo production of the A5H. Although the most capable version of the Atlas V is not designated a "heavy" launcher, its lift capacity places it in the heavy-lift launch category as defined by the Federal Aviation Administration.

[26] Those payloads are assumed to be lifted into an orbit of 220 nautical miles in altitude inclined at 28.6 degrees. The combined weight of the space shuttle orbiter and the payload that is lifted into LEO is about 100 mt; however, the orbiter returns to Earth after delivering an 18 mt maximum payload into that reference orbit.

[27] That rate attributes the loss of *Columbia* to launch failure because its breakup during reentry resulted from damage caused during liftoff by the shedding of foam insulation from the external tank.

[28] When computed using a large sample of launches, there is no reason to expect the reliability of the Delta IV Heavy to be any better or worse than the demonstrated reliability of similar expendable launchers. However, the D4H has flown only once, and it failed to place its test payload into the intended orbit. The problem—cavitation in a fuel line, which could affect all Delta IVs—was discovered and fixed. The Delta IV Medium resumed flying on May 24, 2006.

[29] The close-derivative launchers considered by CBO are based on information received from contractors and NASA from September 2004 through October 2005.

[31] That assessment is based on the contractor's estimates of the vehicles' reliability.

[32] The U.S. Air Force formally projects in its future manifests that it will fly about nine EELV flights annually, split roughly equally between the Delta IV and Atlas V. The annual number of EELV flights conducted recently, however, has been less than one-third of that projection.

[33] Builders of launch vehicles usually regard the development of new engines as the riskiest aspect of constructing new launchers. A risk-reduction step used by NASA and industry teams is to rely on proven engines (or only slight variations) but to build larger launchers to house more of them.

[34] Ignoring the unavailability of parts and suppliers, reconfiguration might be necessary because the Saturn V does not meet current NASA safety standards (see Appendix B).

[35] See Office of Management and Budget, *Budget of the United States Government, Fiscal Year 2007* (January 2006).

[34] In May 2006, NASA announced that it was changing the design of its super heavy to incorporate key features of both the shuttle- derived super heavy and the Longfellow. The new design for the 130 mt cargo carrier replaces the five SSMEs powering the vehicle's first stage with five of the RS-68 engines used on the first stage of the Longfellow. (Unlike the Longfellow, the newly designed cargo carrier would not have a powered second stage.) Five RS-68 engines would require more space under the tank, which in turn would require that the diameter of the first stage be larger than that of the existing ET. CBO lacks the information needed to prepare other than a preliminary estimate of the costs of this new cargo-carrier design.

[36] CBO's preliminary estimates of the costs of NASA's latest design for the shuttle-derived super-heavy cargo carrier using five RS-68 engines indicate that the lower bound on those recurring costs could be about $1.0 billion and that the lower bound on nonrecurring costs could be about $7.4 billion.

[37] Details of the infrastructure requirements and their relatively high expense were provided to CBO on August 5, 2005, during discussions with members of Boeing's Launch Services Division in Huntington Beach, Calif.

[38] Based in part on the *Columbia Accident Investigation Board Report* and NASA's decision to separate cargo and crew launches, CBO did not consider a case in which one Longfellow

would launch cargo and a second would launch the remaining cargo, as well as the lunar spacecraft and its crew.

[39] CBO's preliminary estimate of the costs of NASA's latest design for the shuttle-derived super heavy, which would use five RS-68 engines instead of SSMEs, suggests that this relationship could be reversed: rather than 10 percent lower, the unit cost of the Atlas super-heavy cargo carrier could be about 20 percent higher.

[40] Reliability calculations are complex, involving a number of tradeoffs. On the one hand, more engines and sticks imply the need for more tanks, valves, sensors, and moving parts, all of which present more failure modes. On the other hand, with enough engines, if one failed to ignite (one of the most common failure modes for engines), the remaining engines might still be able to deliver a payload into its intended orbit.

[41] The first stage of the Soviet N-1 moon rocket had 30 engines. All four test flights of that rocket ended in failure before separation of the first stage could be completed.

[42] NASA also supports the ISS with funds in another account called the ISS Cargo and Crew Services (ISS CCS) project. This account funds both purchases of launches from foreign providers (in particular, Russia) and development of U.S. commercial launch services to LEO. In NASA's budget projections through 2017, which were prepared earlier this year, the agency assumes that its crew carrier will be used for five ISS support flights annually, with the new ISS CCS project providing one additional support flight to the ISS annually.

[43] An operational launch is executed in order to conduct a mission to the space station or to the moon. CBO distinguishes those launches from the test launches conducted in order to assure that a newly developed launcher is ready for operational use. The number of test launches conducted through 2017 under any given alternative is the same for either the more ambitious or less ambitious schedule.

[44] See Office of Management and Budget, *Budget of the United States Government, Fiscal Year 2007*.

[45] CBO's preliminary estimate for the costs of NASA's latest design for the shuttle-derived super-heavy cargo carrier, which uses five RS-68 engines, indicates that programmatic costs for the more ambitious schedule could be about $30 billion, essentially the same amount contained in NASA's future funding projection.

[46] CBO's preliminary estimate for the costs of NASA's latest design for the shuttle-derived super-heavy cargo carrier, which uses five RS-68 engines, indicates that programmatic costs for the less ambitious schedule could be about $19 billion, which is about $11 billion less through 2017 than NASA's future funding projection.

[47] NASA, *Overview of the DART Mishap Investigation Results*, available at www.nasa.gov/pdf/148072main_DART_mishap_ overview.pdf.

[48] The "eyeballs in" direction assumes that the crewmembers reenter the atmosphere oriented so that their backs face the direction of motion. Other orientations also are important, but the eyeballs in direction is used here to make comparisons.

[49] Oxygen and other gases also leak but not as extensively, which is why hydrogen leakage usually dominates considerations.

[50] Actual boil-off rates are complex functions of the material used to construct the storage container, its geometry, and whether and how passive and/or active cooling techniques are employed. A boil-off rate of 34 percent is typical for passive cooling techniques and traditional storage containers.

[51] Data extracted from the U.S. Air Force's Eastern Range Launch Performance, 1988 to 2002.

[52] The Air Force's policy of assured access is a justification for both the Atlas V and Delta IV lines of launchers. In the absence of Vision for Space Exploration missions, either one of those lines has the infrastructure in place to meet (or, in 2012, to nearly meet) the Air Force's launch demands to place payloads of up to about 20 metric tons in orbit. Peak demands are taken from an Air Force manifest of projected launch demand through 2020. However, the recent flight history of the EELV program suggests that actual flights may be significantly lower than projections.

[53] The safety and human aids collectively fall under NASA's launch vehicle health management (LVHM) system.

[54] The burden rate assumes full cost accounting and represents government overhead associated with the staffing and maintenance of NASA's facilities.

In: Future U.S. Space Launch Capabilities
Editors: Franz Lojdahl
ISBN: 978-1-60741-384-4

Chapter 2

Space Acquisitions: Uncertainties in the Evolved Expendable Launch Vehicle Program Pose Management and Oversight Challenges

Government Accountability Office

Why GAO Did This Study

The Department of Defense (DOD) plans to spend over $27 billion acquiring launch services through the Evolved Expendable Launch Vehicle (EELV) program over the next 12 years. The EELV program uses two families of commercially owned and operated vehicles to launch satellites.

Partly because the commercial space market did not develop as expected, the EELV program has undergone significant changes. These include: adoption of a new acquisition strategy in 2005 that sought to ensure the viability of the two EELV launch vehicle providers, Boeing and Lockheed Martin; the subsequent decision by those two companies to form a joint venture called the United Launch Alliance (ULA); and a 10- year increase in the life of the program. In light of these changes, GAO was asked to (1) determine what uncertainties DOD faces in the EELV program and in the transition to ULA, and (2) assess how DOD is positioned to manage and oversee the effort. To accomplish this, GAO reviewed a wide variety of DOD documents and interviewed DOD and program officials.

What GAO Found

The EELV program currently faces uncertainties in the reliability of the vehicles used to launch military and other government spacecraft as well as its budget for future years and in the merger of its two principal suppliers. Taken together, these unknowns require careful monitoring and oversight to ensure a fairly long track record of launch successes can continue.

- Though the program has had 21 successful operational launches, no single configuration from either family of launch vehicles has been launched enough times to demonstrate production process reliability. The ULA transition may also influence the demonstration of vehicle reliability because ULA plans to relocate production activities and may alter manufacturing processes.
- The consolidation of Boeing and Lockheed into ULA—a massive undertaking that seeks to combine two distinct corporate cultures, and consolidate launch infrastructure and business operations from five locations across the country to two—poses a variety of other cost, schedule, and performance uncertainties and risks.
- DOD does not know whether its EELV program budget is sufficient to manage the program in the short term because the Air Force reduced the EELV program budget to incorporate anticipated savings from the ULA transition, even though savings estimates were based on preliminary data.

DOD has taken steps to position itself to effectively oversee and manage the ULA transition and EELV program but still faces significant challenges in these areas. More specifically, DOD has established a well-defined process for how the ULA transition will be overseen and established mechanisms that allow the diverse agencies involved to coordinate the analysis and raise critical issues to senior leaders. However, when DOD moved the EELV program from the research and development phases in 2007 to the sustainment phase, DOD eliminated requirements on the program to produce data that would illuminate what impacts the transition is having on the program, what cost increases are occurring and why, and what other programmatic and technical vulnerabilities exist and how they are being addressed. Furthermore, a new independent life-cycle cost estimate was not required for the program when it moved to the sustainment phase; as a result, DOD will not be able to rely on this estimate for making long-term investment planning decisions. According to DOD officials, the latest life-cycle cost estimate for the program is not realistic. In addition, as part of its effort to increase its oversight and gain program knowledge, in 2005 the Under Secretary of the Air Force expanded the program office's management responsibilities when he approved a new acquisition strategy for the EELV program. At the same time, program officials stated that they do not have the government staff necessary to perform what they consider to be inherently governmental functions related to the expansion of oversight.

ABBREVIATIONS

AFB	Air Force Base
CAE	component acquisition executive
DAES	Defense Acquisition Executive Summary
DCAA	Defense Contract Audit Agency
DCMA	Defense Contract Management Agency
DOD	Department of Defense
EELV	Evolved Expendable Launch Vehicle
EVM	earned value management
FAR	Federal Acquisition Regulation
FFRDC	federally funded research and development center
FTC	Federal Trade Commission
FTE	full-time equivalent
NASA	National Aeronautics and Space Administration
NRO	National Reconnaissance Office
NSSO	National Security Space Office
OSD	Office of the Secretary of Defense
PBD	Program Budget Decision
PEO	program executive officer
SAR	Selected Acquisition Report
SETA	systems engineering and technical assistance
SMC	Space and Missile Systems Center
ULA	United Launch Alliance

The Department of Defense (DOD) plans to spend over $27 billion acquiring launch services through the Evolved Expendable Launch Vehicle (EELV) program over the next 12 years. DOD's investment in the EELV program, which uses two families of commercially owned and operated vehicles to launch satellites, is intended to reduce the overall cost of space launches, assure access to space, and preserve the space launch industrial base.

The EELV program has undergone significant changes since its inception. DOD's initial acquisition strategy was to have multiple companies develop launch vehicles and then pick one launch vehicle provider in 1998. However, in November 1997 DOD decided to keep the two contractors then competing, Boeing and Lockheed Martin, as EELV providers, and later awarded contracts to both companies. DOD officials had expected the emergence of a healthy commercial launch market that would support two families of launch vehicles and ultimately mean lower costs to the government. DOD determined launch service prices in the contracts based on an assumption that a robust commercial market would help pay for the launch providers' infrastructure costs. However, in the late 1990s the commercial launch market collapsed, and in late 2003 DOD estimated that average launch cost would increase significantly. DOD responded to these developments with a new acquisition strategy in March 2005 that recognized the government as the primary customer of EELV services and sought to ensure the viability of the two EELV launch vehicle providers.

Additionally, in December 2006 Boeing and Lockheed Martin formed a joint venture called the United Launch Alliance (ULA) to provide EELV launch services to the government. The ULA intends to maintain two separate launch vehicle families while consolidating business, engineering, and manufacturing activities to reduce costs. This consolidation entails moving launch vehicle production facilities and work forces to new locations. Finally, another significant change is a recent 10-year extension to the life of the EELV program.

In light of the changes the EELV program has experienced, you asked us to (1) determine what uncertainties DOD faces in the EELV program and in the transition to the United Launch Alliance, and (2) assess how DOD is positioned to manage and oversee the program.

To determine what uncertainties DOD faces in the EELV program and ULA transition, we reviewed EELV budget and program status documents, studies of EELV reliability and launch history, ULA joint venture transition plans and status, and DOD risk tracking and mitigation documentation. To assess how DOD is positioned to manage and oversee the EELV program, we analyzed data on EELV program office[1] authorized staffing levels and compared them to filled positions. We also reviewed reporting requirements under the new EELV acquisition strategy, contract data requirements lists for both Atlas and Delta contracts, EELV program oversight documentation, and DOD acquisition policies. We interviewed officials from various DOD organizations as well as launch contractors. We conducted this performance audit between June 2007 and September 2008 in accordance with generally accepted government auditing standards. Those standards require that we plan and perform the audit to obtain sufficient, appropriate evidence to provide a reasonable basis for our findings and conclusions based on our audit objectives. We believe that the evidence obtained provides a reasonable basis for our findings and conclusions based on our audit objectives.

RESULTS IN BRIEF

DOD faces numerous uncertainties in the EELV program and ULA transition related to the reliability of the launch vehicles, the amount of work remaining in the ULA transition, and program budget decisions based on preliminary data. First, though the program has had 21 successful operational launches, DOD does not yet know whether EELVs are reliable because no single configuration from either family of launch vehicles has been launched at least 7 times to demonstrate production process reliability. According to a widely accepted measurement practice, production process maturity can be demonstrated through 7 successful launches. Additionally, DOD may be less certain of EELV reliability as the ULA transition progresses, because ULA plans to relocate production activities and may alter manufacturing processes. Second, DOD faces uncertainties relating to the cost, schedule, and performance risks of the ULA transition, mainly because the transition is still in its initial stages, and numerous unknowns exist in areas such as production, personnel management, and cost savings. Furthermore, the ULA consolidation is a massive undertaking as it attempts to combine two distinct corporate cultures into a unified workforce, and consolidate launch infrastructure and business operations from five locations across the country to two. Third, DOD does not know whether its EELV program budget is sufficient to manage the program in the short term because the Air Force reduced the EELV program budget to incorporate anticipated savings from the ULA transition, even though savings estimates were based on preliminary data.

DOD has taken steps to position itself to effectively oversee and manage the ULA transition and EELV program but still faces significant challenges in these areas. More specifically, DOD has a well-defined process for how the ULA transition will be overseen and established mechanisms that allow the diverse agencies with oversight responsibilities to coordinate the analysis of the transition and raise critical issues to senior leaders. However, by moving the EELV program in 2007 to the sustainment phase, as opposed to the research and development and production phases, DOD eliminated requirements on the program to submit data to DOD offices responsible for program oversight—data which could illuminate program variances and technical risks and how they are being addressed. Furthermore a new independent life-cycle cost estimate was not required for the program when DOD moved it from the research and development phases to the sustainment phase. As a result, DOD will not be able to depend on a reliable life-cycle cost estimate for making long-term investment planning decisions because, according to DOD officials, the latest life-cycle cost estimate for the program is not realistic. In addition, as part of its effort to increase its oversight and gain program knowledge, in 2005 the Under Secretary of the Air Force expanded the program office's management responsibilities when it approved a new acquisition strategy for the EELV program. However, program officials stated that they do not have the government staff necessary to perform these expanded duties, which include activities such negotiating contracts, performing earned value management reviews, determining award fees, and assessing contractor performance.

To improve DOD's ability to effectively oversee and manage the EELV program, we are recommending that DOD (1) continue to produce and share data with senior leaders on program cost and status using criteria that apply to major research and development and procurement programs and (2) prepare an independent life-cycle cost estimate once the ULA joint venture transition is completed so that senior leaders and the program can make decisions based on an estimate that reflects the span of changes that have occurred in recent years. We are also recommending that the Air Force assess the EELV program's staffing needs to determine the extent to which personnel shortages exist, particularly as they relate to functions integral to achieving DOD's goals for enhanced program oversight. DOD agreed with our recommendations.

BACKGROUND

DOD initiated the EELV program in 1995 to develop a new generation of launch vehicles. The EELV program— designed to provide assured, affordable access to space for government satellites—consists of two families of commercially owned and operated launch vehicles, the Atlas V and Delta

IV launch vehicles. It also includes manufacturing and launch site facilities, and ground support systems. Each family of launch vehicles consists of medium-, intermediate-, and heavy-lift vehicles. To simplify the design and manufacture of the vehicles, common components (such as booster engines) are being extensively used. See figures 1 and 2 for depictions of the Atlas V and Delta IV launch vehicle families and major components of the launch vehicles.

In 1995, the Air Force awarded $30 million firm fixed-price contracts to four companies to define EELV system concepts and complete preliminary system designs. In December 1996, two of those companies, McDonnell Douglas (which became part of Boeing in 1997) and Lockheed Martin, were each awarded an 18-month, $60 million cost-plus fixed-fee contract to continue EELV design activities. At the end of their contracts, DOD planned to down-select to one contractor with the most reliable and cost- effective launch vehicle design. However, in November 1997 the Office of the Secretary of Defense (OSD) approved maintaining competition between the two contractors. This decision was based on forecasts that growth in the commercial space launch services market would support more than one launch provider and translate into lower costs for the government. More specifically, the strategy was designed to

- promote competition for launch services over the life of the program;
- encourage contractor investment and innovation for launch vehicle development;
- procure launch services that include the vehicle, the lift-off, and flight to orbit under one contract at a fixed price instead of procuring launch vehicles and launch operations under two or more separate contracts; and
- leverage the benefits of a robust commercial marketplace.

In October 1998, the Air Force competitively awarded two $500 million "other transaction agreements" to McDonnell Douglas before it merged with Boeing and Lockheed Martin for the development of EELV families of launch vehicles and launch infrastructure. Other transaction agreements are transactions that are other than contracts, grants, or cooperative agreements that are subject to federal laws and regulations applicable to contracts, grants, or cooperative agreements. The Air Force used other transaction agreements to provide flexibility in accommodating the unique needs of the EELV program and the government. Simultaneous with the Air Force's execution of the other transaction agreements, the Air Force also competitively awarded firm fixed-price contracts, one each to McDonnell Douglas (now Boeing) and Lockheed Martin, to procure 28 launches. At the time of award, the launch services contracts had a combined value of $2.03 billion.

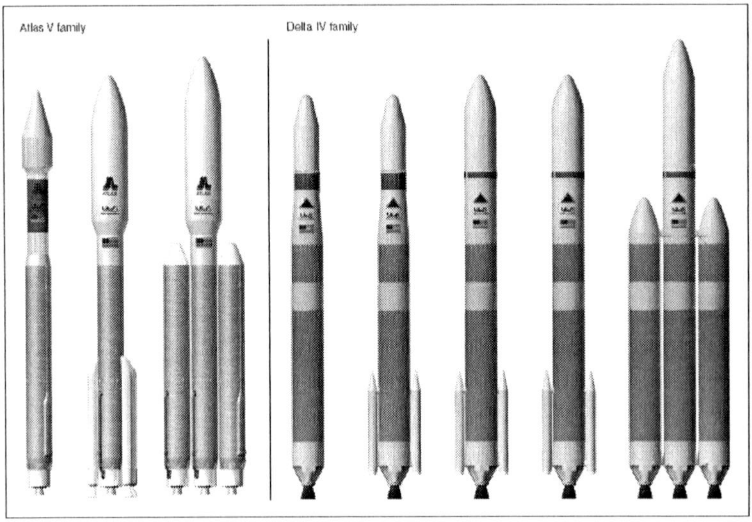

Note: The Atlas V heavy-lift vehicle has not been built.

Figure 1. Atlas V and Delta IV Families of Launch Vehicles

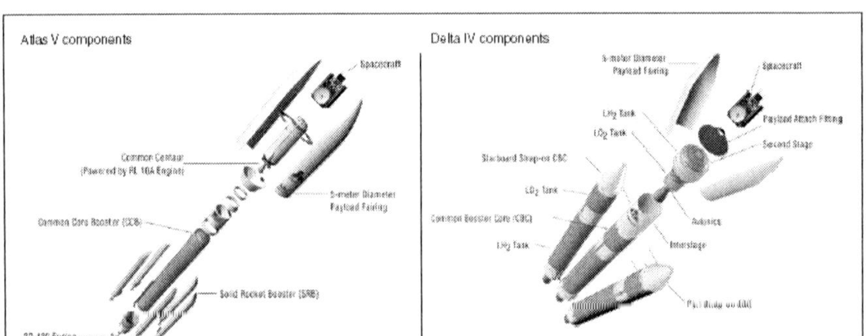

Figure 2. Major Components of Atlas V and Delta IV Launch Vehicles

The robust commercial launch market upon which OSD based the EELV acquisition strategy did not materialize. As a consequence, estimated prices for future contracts for launch services increased, along with the total cost of the program. DOD's December 2003 cost estimate for the program showed an average procurement unit cost for EELV launch services that was 77 percent higher than its 2002 cost estimate. This resulted in a Nunn-McCurdy unit cost breach, which occurs when a major defense acquisition program experiences a unit cost increase of at least 15 or 25 percent.[2] In April 2004, in response to the unit cost breach, the Secretary of Defense sent Congress a certification that the program was essential to national security and established a new cost estimate. The new cost estimate increased the launch service's average procurement unit costs by an additional 29 percent, due primarily to a reduction in the number of planned EELV launches.

In March 2005, DOD revised the EELV acquisition strategy to reflect changes in the commercial market assumptions since 1998 and the role of the government as the primary EELV customer. The revised approach involved a new contracting strategy consisting of two negotiated contracts—EELV Launch Capability and EELV Launch Services—for each of the two launch service providers. The EELV Launch Capability cost-plus award fee contract is primarily for launch infrastructure, while the EELV Launch Services fixed-price incentive fee contract is for launch services. Under this contract approach, the government purchases launch services through competitive negotiated acquisitions that provide more government insight into contractor costs than contracts awarded under commercial item acquisition procedures.[3] This enables the Air Force to collect previously unavailable certified cost or pricing data. As with the original acquisition strategy, there are no government ownership costs— the EELV program continues to provide launch services and ownership of hardware and facilities is retained by the contractors.

In May 2005, Boeing and Lockheed Martin announced plans to create a joint venture that would combine the production, engineering, test, and launch operations associated with U.S. government launches of Boeing's Delta and Lockheed Martin's Atlas rockets. According to both contractors, the joint venture, named the United Launch Alliance, would reduce the cost of meeting critical national security and National Aeronautics and Space Administration (NASA) expendable launch vehicle needs. By joining together, the companies stated they could provide the government with assured access to space at the lowest possible cost.

The ULA was structured as a 50-50 joint venture between Boeing and Lockheed Martin—combining services provided separately by Boeing

Integrated Defense Systems' Expendable Launch Systems division and by Lockheed Martin's Space Systems Company—for launches of each company's respective rockets. Based upon initial contractor estimates, annual savings to the government resulting from the joint venture were expected to be approximately $100 million to $150 million.

The ULA joint venture—which was subject to government and regulatory approval in the United States and internationally—drew initial criticism from various organizations. The Federal Trade Commission (FTC) was initially opposed to the ULA joint venture because of the potential to limit competition in the launch industry—an industry with already high barriers to entry. Other companies in the launch industry also submitted formal concerns citing unfair competition. In response, DOD stated that having two launch vehicle families would present unique national security benefits that would outweigh the loss of competition. FTC agreed to allow the joint venture to move forward under the conditions of a consent order, which requires ULA to be nondiscriminatory to other companies seeking launch services. FTC provided preliminary approval of the consent order in October 2006,[4] and the joint venture officially began operations in December 2006. To ensure that ULA follows the provisions of the consent order, the order requires the Secretary of Defense to appoint a compliance officer, currently the Director of the National Security Space Office (NSSO),[5] to have broad investigative and remedial powers to ensure compliance. In addition, the consent order requires ULA to establish—no later than 24 months after the formation of ULA, or December 2008—separate communication networks and management information systems from its parent companies, Boeing and Lockheed Martin.

The first Atlas V and Delta IV launches occurred in 2002. According to the Air Force, all 21 Atlas V and Delta IV launches through June 2008 have been successful.[6] DOD estimates its investment in the EELV program through fiscal year 2008 to be about $8.2 billion, and future investment to be about $27.5 billion.

DOD Faces Numerous Uncertainties in the EELV Program and ULA Transition

DOD faces numerous uncertainties in the EELV program and ULA transition related to the reliability of the launch vehicles, the amount of work remaining in the ULA transition, and program budget decisions based on premature data. First, DOD does not yet have enough launch experience with EELVs to demonstrate their reliability, which may be accomplished by carrying out a set number of successful launches. While all Atlas V and Delta IV operational launches conducted thus far have been successful, launch vehicle design and production reliability have yet be demonstrated due to the limited number of launches. Additionally, DOD may be even less certain of EELV reliability as the program progresses as plans to relocate production activities under the ULA transition may undo some of the progress made toward demonstrating EELV reliability. Second, DOD faces uncertainties relating to the cost, schedule, and performance risks of the ULA transition, mainly because the transition is still in its initial stages, and numerous unknowns exist in areas such as business and production processes, personnel management, and cost savings. Third, DOD does not know whether its EELV program budget is sufficient to manage the program over the next 5 years because the Air Force accounted for anticipated savings from the ULA transition in the program budget,[7] even though savings estimates were based on preliminary data.

DOD Does Not Yet Know if EELVs Are Reliable

DOD does not yet have enough launch experience with EELVs to demonstrate their reliability. According to the Air Force, since the first launches in 2002, all 21 Atlas V and Delta IV operational launches have been successful.[8] However, these launches represent experience with only 9 of 15 possible EELV configurations across both families of launch vehicles. As depicted in table 1, only 1 EELV configuration has been launched more than 3 times. This is significant because the number of times a configuration has been successfully launched can be an indicator of a configuration's design and production process reliability, according to an Aerospace Corporation[9] study of EELV launch risk.[10]

Table 1. Evolved Expendable Launch Vehicle Configurations by Number of Operational Launches

	Launch vehicle name		Vehicle configuration[a]	Number of operational launches
1	Atlas V	401	4 – 0	6
2		411	4 – 1	2
3		421	4 – 2	2
4		431	4 – 3	1
5		501	5 – 0	0
6		511	5 – 1	0
7		521	5 – 2	2
8		531	5 – 3	0
9		541	5 – 4	0
10		551	5 – 5	1
11	Delta IV	Medium	4 – 0	3
12		Medium+ (4,2)	4 – 2	3
13		Medium+ (5,2)	5 – 2	0
14		Medium+ (5,4)	5 – 4	0
15		Heavy[b]	Heavy	1
	Total			**21**

Source: GAO analysis of Air Force data.

[a] The first number refers to the diameter (in meters) of the compartment on the launch vehicle that houses the payload and the second number refers to the number of add-on solid rocket boosters mounted to the primary booster. The heavy configuration refers to a launch vehicle consisting of three primary boosters. See figures 1 and 2 for depictions of the launch vehicles and their components.

[b] Does not include a 2004 Delta IV heavy demonstration launch which did not deliver the payload to the proper orbit.

Through a study of the causes of worldwide launch vehicle success and failures over a 13-year period, the Aerospace Corporation developed a widely accepted measurement for demonstrating launch vehicle design and production process reliability called the "3/7 Reliability Rule." According to this rule, if a failure occurs during a configuration's first three launches, the problem is most likely a design issue, and if a failure occurs after the third successful launch but before the seventh, the problem is most likely a production process issue. Once a launch vehicle configuration is launched successfully three times, its design can be considered to have demonstrated maturity; similarly, if a vehicle is successfully launched seven times, both the design and production process maturity of that launch vehicle configuration can be considered as having been demonstrated. Until a launch vehicle configuration demonstrates maturity, problems encountered during launches of these configurations are more likely to uncover fleetwide design or production process issues that could cause significant cost and schedule impacts to the program.

Table 1 shows that only 3 of the 15 EELV configurations have been launched three or more times, and none has been launched seven times. The EELV program office agrees that the 3/7 Reliability Rule is applicable to EELV configurations, and program officials state that the program will continue to apply the rule when assessing the requisite level of quality assurance activities for each configuration. However, according to OSD officials, the 3/7 Reliability Rule is only partially applicable to EELV configurations because of the commonality among the vehicles. For example, the same common core boosters are used on all Atlas vehicles; therefore, according to these officials, reliability on the Atlas launch vehicle line can be demonstrated through fewer launches of individual configurations than the 3/7 Reliability Rule dictates. The same is true of Delta launch vehicles. Nevertheless, these officials agreed that inherent risk lies in the integration of launch vehicle configurations, which can be lessened with each consecutive flight.

ULA Transition Activities May Undermine Progress Made in Demonstrating EELV Reliability

Transition activities under the ULA joint venture include moving production equipment to different facilities and possibly changing manufacturing processes, and could reverse some of the progress made toward demonstrating production process maturity of those EELV configurations launched previously. Specifically, part of the ULA transition plan includes relocating production of the Atlas line of launch vehicles from San Diego, California, and Denver, Colorado, to Decatur, Alabama. According to Aerospace Corporation officials and a 2007 study by Booz Allen Hamilton,[11] relocating highly calibrated production equipment from one facility to another and changing manufacturing processes increases risk to launch vehicle quality and may detract from progress made toward demonstrating production process maturity. This means that launch vehicle configurations (shown in table 1) that have been successfully launched, and have therefore begun the process of establishing production process maturity, may lose some of their accumulated progress.

The ULA transition plan also allows for changing one of its key manufacturing processes. While the process may be more effective, it has not been used before in Atlas production, though it is currently used in the production of Delta launch vehicles. According to ULA officials, incorporating the new process on Atlas vehicles could potentially reduce the need for some rework. Risk reduction testing through July 2008 has shown the technical viability of the new process, according to the EELV program office, and additional discussions with ULA will determine whether ULA will pursue it on the Atlas line of vehicles.

Given the limited EELV launches to date, and the possibility that some progress toward demonstrating maturity may be lost due to relocation of production or changes to the process, DOD can not be fully confident in the design and production process maturity of most of the EELV configurations. According to program officials, although some EELV configurations may never demonstrate design or production process maturity throughout the life of the EELV program because of their limited demand, using data from launches of other configurations provides insight into the reliability estimates for those vehicles that have not yet flown.

DOD Faces Numerous Uncertainties about the Cost, Schedule, and Performance Risks of the ULA Transition

Any corporate consolidation on the scale of the ULA joint venture poses challenges and a great deal of uncertainty. In addition to bringing together diverse cultures and ways of doing business, such endeavors may involve complicated personnel relocations, the closing and opening of facilities, implementation of new information technology systems, and rethinking of business processes, while at the same time maintaining high-quality customer service. Because many ULA transition activities have only recently begun or are still in the planning stages, numerous uncertainties exist, many of which are inherent in the transition itself. They span business and production operations, personnel management, and cost savings, and have the potential to significantly affect the cost of the transition, and ultimately, launch vehicle quality. Following are a few examples of these uncertainties.

- In addition to possibly introducing a new manufacturing process as stated earlier, ULA will incorporate Atlas manufacturing equipment into the Decatur facility. Current plans include a physical reconfiguration of the facility to accommodate the influx of Atlas production, including the movement of interior walls and a change in production flow for the Delta production line, both of which could affect launch vehicle quality.
- As Atlas production operations move to Decatur, ULA will try to relocate as many Atlas staff as possible. ULA's goal is to move roughly 30 percent of staff from San Diego and Denver to Decatur, but Defense Contract Management Agency (DCMA) officials state that they expect it will be a challenge for ULA to retain even 10 percent of targeted Atlas staff. The loss of manufacturing knowledge likely to result from the relocation of Atlas production operations is regarded by DOD, Aerospace Corporation, and ULA officials as one of the top risks of the transition because the loss of critical manufacturing skills and process knowledge could ultimately affect launch vehicle quality. The relocation of Delta engineering and program operations from Huntington Beach, California, to Denver as part of the ULA consolidation may also result in the loss of critical engineering skills, according to Aerospace Corporation officials, as many experienced staff will not remain under ULA employment. ULA offered positions in Denver to all of its salaried employees in the Huntington Beach facility, and 41 percent reportedly accepted the offer to relocate.
- Efforts to separate information technology and management networks from parent companies and to set up accounting practices are proving more complex and time consuming than originally anticipated and a recent 10-month extension to the goal date of separation is expected to result in higher-than-anticipated costs. As it moves toward separation, ULA is buying infrastructure support services from its parent companies, through "transition service agreements," and the additional costs associated with prolonging the use of these agreements could ultimately increase the cost of the separation effort. Furthermore, financial repercussions are possible due to the large amount and complexity of the work remaining to accomplish the FTC-mandated information systems separation. The ULA recently requested a formal extension to the FTC's specified deadline of December 2008, which was granted by the NSSO Compliance

Officer. The revised deadline for separation is October 2009, representing a 10-month delay.

- ULA has provided DOD with a restructure proposal which projects the cost and savings of the transition. According to DCMA, as a result of review, the proposal currently meets federal statutory requirements for the reimbursement of restructuring expenses. An advance agreement that accompanies the proposal specifies the maximum amount of reimbursement ULA may recover from DOD for restructuring expenses. DOD is currently reviewing the proposal and advance agreement, in part, to determine if they are based on sound business judgment and are equitable to all parties. However, much of the transition has yet to occur, and some of the transition activities that have begun are proving more complex than anticipated. ULA estimates that consolidation of launch infrastructure and operations under the ULA joint venture from fiscal years 2008 through 2012 will cost about $205 million, and it plans to recover much of this expense from the government. According to federal statute, the government may not reimburse a contractor—in this case ULA—for the costs of the restructure, unless DOD determines in writing allowed, that (1) the savings to DOD are at least twice the amount of the costs allowed, or (2) the amount of projected savings to DOD will exceed the amount of the costs allowed and that the business combination will result in the preservation of a critical capability that otherwise might be lost to the department.[12] Should the cost of transition activities exceed ULA's estimate, then according to DCMA, ULA may face pressure to meet the requirements for reimbursement of restructure costs. We agree with DCMA that if this happens, ULA may be at risk of making decisions based more on cost considerations than on assuring quality within the launch vehicle manufacturing process.

Table 2 provides a more complete description of ULA transition uncertainties.

DOD Adjusted the EELV Program Budget Using Premature Savings Estimates

As described in table 2, despite the magnitude of activities that have yet to be completed under the ULA transition, DOD has accounted for over $100 million in savings in the EELV program budget, beginning in 2011. In 2005, before the joint venture was finalized, a small contractor team that was formed to review the proposed joint venture estimated the likely implementation costs and projected the savings that would result. Preliminary calculations showed that the consolidation of launch infrastructure and operations would yield up to $150 million in savings per year, beginning in 2011. The Air Force portion of these savings was projected at $105 million per year.[13] Program officials stated that while formulating the program budget in 2006, prior even to final government approval of the ULA joint venture, the Air Force was confident enough in the estimates to subtract $105 million in savings per year from the EELV program budget, to begin in 2011. Given the many activities remaining in the transition, and the preliminary nature of the data used to calculate the savings, it may have been premature for DOD to reflect the anticipated savings in the EELV program budget. Furthermore, some transition activities that have begun are expected to take longer than originally planned, as discussed above. DOD officials stated that ULA's current proposed savings estimate undergoing DOD review shows greater annual savings than originally estimated. However, if the cost or duration of the ULA transition is different than anticipated, actual savings could differ from the predicted $105 million per year beginning in 2011, leaving DOD at risk of EELV program budget shortfalls.

DOD Has Made Strides to Effectively Oversee and Manage the EELV Program but Challenges Exist in These Areas

In recent years, DOD has taken steps to effectively oversee and manage the EELV program by closely monitoring the ULA transition and changing its strategy for acquiring launch services. However, OSD's decision to advance the program into the sustainment phase is limiting its ability to provide effective oversight. In addition, the EELV program office believes a shortage of staff with the right skills and experience is limiting its ability to effectively manage the program. Air Force staffing initiatives have done little to alleviate this shortage.

Table 2. Uncertainties of the ULA Transition

ULA uncertainties	Description
Business and production uncertainties	• Implementation of information technology systems has experienced multiple delays, and actual implementation time frames are unknown.
	◦ Implementation of the Enterprise Resource Planning system is taking significantly longer than expected, and DCMA expects deployment of this system to cost nearly twice as much as originally estimated.
	◦ According to DCMA, milestones for implementing the Manufacturing Execution System, a part of the Enterprise Resource Planning system, have slipped as much as 1 year because ULA's original schedule was too aggressive; ULA agrees schedule was inadequate.
	• Several company stand-up activities have experienced delays and as of September 2008 approval is still pending for
	◦ disclosure statement amendments outlining new accounting practices and rate structures, and
	◦ a restructure proposal which finalizes the transition plan and resulting cost savings.
	• Separation of information technology and management networks from parent companies is taking longer than planned, and has required a 10-month extension to FTC's December 2008 deadline.
	• Relocating Atlas production and reconfiguring Delta equipment may reverse some of the progress made toward establishing production process reliability.
	• Modifications to the Decatur facility's building to accommodate Atlas production could disrupt current processes for Delta production, resulting in potential loss of progress toward demonstrating production process reliability.
	• Changing one of the key manufacturing processes on the Atlas family of launch vehicles may reverse some of the progress made toward establishing production process reliability.

Table 2. (Continued)

ULA uncertainties	Description
Personnel management uncertainties	• Loss of critical Atlas manufacturing and process knowledge could result from inadequate personnel relocation and could impact launch vehicle quality.
	• Loss of engineering knowledge could result from impending separation of senior Delta engineering staff from ULA, in turn impacting launch vehicle quality.
	• Combining two distinct corporate cultures while maintaining a motivated, workforce may be difficult and morale could suffer.
Cost savings uncertainties	• Many ULA activities have yet to begin; if ULA underestimated resources needed for these activities, costs could increase.
	• Extending the FTC deadline for ULA separation of information technology and management networks from parent companies could result in prolonged need to purchase infrastructure support services from parents, potentially increasing costs.
	• Additional training costs could arise if ULA does not retain adequate personnel following relocation of Atlas production.
	• According to DCMA, if ULA does not pursue the new manufacturing process on Atlas vehicles as originally planned, ULA will have to replace the equipment currently used, potentially increasing cost, as relocating this equipment to the Decatur facility is not an option.
	• The EELV program office and DCMA are evaluating the qualification requirements for the first Centaur tank produced following relocating production to Decatur. One option could entail the production of a tank solely for testing purposes, and if this option is executed, it would deviate from ULA's baseline, which includes testing a tank intended for operational use. This means production of the testing tank and the associated testing regimen could increase costs.

Source: GAO analysis of data from DCAA, DCMA, the EELV program office, and ULA.

Table 3. DOD Responsibilities in Monitoring the ULA transition

Organization	Responsibilities
National Security Space Office (NSSO)	• Verifies compliance with the FTC decision and order
	• Meets periodically with ULA to determine progress made
	• Reviews written reports by Lockheed Martin, Boeing, and ULA showing progress made in achieving compliance with the FTC consent order
	• Reports conclusions to the FTC annually
Defense Contract Management Agency (DCMA)	• Performs the majority of transition monitoring efforts, conducts surveill-ance through communications with ULA, conducts risk assessments, and documents and reports results of its evaluations to the program office
	• Participates in periodic Executive Board meetings to review progress of transition
	• Provides status updates and raises critical issues to DCMA Headquarters for guidance
	• Set up "Red Teams" consisting of DCMA officials at ULA facilities in Decatur, Alabama, San Diego, California, and Denver, Colorado, to monitor transition operations
	• Makes determinations and recommendations including, but not limited to, indemnification packages, cost accounting system acceptability, novation[a] agreements, and restructure proposal
EELV Program Office	• Reviews documents such as indemnification packages, novation agreements, and the restructure proposal and coordinates with DCMA and other appropriate offices
	• Holds regular meetings with ULA to obtain updates of transition
	• Provides progress updates to Air Force leadership in Executive Board meetings
	• Monitors, with Aerospace Corporation assistance, the risks of the ULA transition to launch success
Defense Contract Audit Agency (DCAA)	• Supports DCMA and the program office by reviewing and providing audit findings on ULA proposals
Office of the Under Secret-ary of Defense for Acquisitions, Technology and Logistics[b]	• Approves final restructure proposal

Source: GAO presentation of DOD data from DCAA, DCMA, EELV program office, and NSSO.

[a] With respect to government contracts, a novation agreement is a legal instrument executed by the contractor (transferor), successor in interest (transferee), and government, and by which, among other things, the transferor guarantees performance of the contract, the transferee assumes all obligations under the contract, and the government recognizes the transfer of the contract and related assets. FAR 2.101.

[b] The Under Secretary of Defense for Acquisitions, Technology and Logistics serves as an advisor to the Secretary of Defense for all matters involving acquisitions, technology, and logistics.

Multiple Offices within DOD Are Monitoring the ULA Transition

DOD has taken positive steps to oversee the ULA transition. Various organizations are monitoring the transition efforts and providing input to OSD, the Air Force, and ULA on ways to mitigate risk. These efforts range from activities by high-level OSD offices such as the Under Secretary of Defense for Acquisitions, Technology and Logistics, down to the day-to- day monitoring activities coordinated by the EELV program office. For example, the Secretary of Defense assigned the Director of the National Security Space Office as the FTC consent order compliance officer to meet periodically with ULA to determine progress made and formally report to FTC on an annual basis, while the EELV program office holds regular meetings with ULA and provides progress updates to Air Force leadership. DCMA conducts ongoing technical risk evaluations, and advises the program office and ULA on a consistent basis of relevant risk reduction measures. As table 3 shows, multiple organizations coordinate with each other to monitor the transition.

DOD officials stated that frequent coordination and communication between the contractor and the government and among the various offices within DOD have ensured a good understanding of progress and issues to date. For example, according to a DCMA senior official, DCMA—which is responsible for many of the transition monitoring efforts—actively participates in Executive Board meetings where all issues pertaining to the transition are raised. The Executive Board consists of senior representatives of several organizations and is chaired by the Air Force Program Executive Officer for Space, who is responsible for all acquisition programs at the Air Force Space Command's Space and Missile Systems Center. The EELV program office, DCMA, NASA, DCAA, National Reconnaissance Office (NRO), and the consent order compliance team are all members of the Executive Board, which convenes approximately quarterly to discuss transition issues, discuss major obstacles, and collectively resolve them. For example, at an Executive Board meeting earlier this year, it was noted that the ULA transition efforts required a detailed master schedule in order to review all schedules for each critical work stream. Consequently, the transition plans now include a much more detailed master schedule that has made it easier to track all critical work activities.

Table 4. OSD Program Cost and Status Reporting Requirements Not Applicable to Programs in the Sustainment Phase

Reporting requirement	Description	Primary user
Defense Acquisition Executive Summary (DAES)	At a minimum, the DAES should report program assessments, unit costs, and current estimates. It should also report the status of exit criteria and vulnerability assessments.	OSD
Nunn-McCurdy unit cost tracking provisions	The Secretary of Defense is required to notify Congress when a major defense acquisition program exceeds certain thresholds.	Congress
Updated program life-cycle cost estimate	Contains summary of program office cost estimate and Cost Analysis Improvement Group independent cost estimate and compares or reconciles the two estimates, includes assessment of program risks, compares (time-phased) Cost Analysis Improvement Group cost estimate to current program funding, and makes recommendations concerning program funding.	OSD and program office
Earned value management (EVM) data	Contains pertinent data that combine measurements of contractor technical performance, such as accomplishment of planned work, schedule performance, and cost performance.	OSD and program office
Selected Acquisition Report (SAR)	Submitted to Congress, contains details on major defense acquisition program cost, schedule, and performance changes on a periodic basis, summarizing the latest estimates of a program's cost, schedule, and technical status.	Congress

Source: GAO presentation of DOD data.

OSD's Capability to Provide Effective Oversight Hampered by Sustainment Decision

While DOD has taken positive steps to oversee and manage the ULA transition, another action within DOD will make it more difficult to monitor the ongoing progress of the EELV program. OSD's decision to advance the EELV program from the development and production phases, which began in 1998, to the sustainment phase will significantly reduce EELV's reporting requirements to OSD, such as program cost and status information, limiting its own ability to oversee the program.[14] According to OSD, the program was placed in sustainment in August 2007 because OSD had determined EELV had completed its development and production phases. However, according to OSD officials, the sustainment decision may have been influenced by other factors such as avoiding imminent Nunn-McCurdy unit cost breaches and possible subsequent certification efforts that invariably would have led to decisions to continue DOD's investment in the EELV program. According to these officials, in 2006 the program was facing one or more possible Nunn-McCurdy unit cost breaches, as a result of unplanned cost increases resulting from the change in the acquisition strategy and a decrease in the demand for EELV launches (factors include satellite program development delays and cancellations, and operational satellites lasting longer than anticipated). These officials stated that even given the anticipated cost savings associated with the ULA joint venture, the program would still have faced a unit cost breach.

When the program was placed into sustainment, however, certain program cost and status information was no longer required because sustainment programs, having completed their development and production phases, generally do not generate the types and kinds of information traditionally produced during development or production. Since this information is generally not produced, program offices are not required to report it to OSD or Congress. For example, the EELV program will no longer be required to submit to OSD a quarterly cost and status report called the Defense Acquisition Executive Summary (DAES). The DAES is a detailed, multipart document that serves as an early-warning report to OSD by describing actual program problems, or warning of potential program problems, and describing mitigating actions taken or planned. According to DOD guidance, the DAES report contains the minimum necessary information for oversight purposes. (See app. II for a more extensive list of what the DAES contains.) Table 4 lists other OSD oversight reporting requirements that do not apply to sustainment programs. According to officials in the program office and the office of the

Assistant Secretary of Defense for Networks and Information Integration, the EELV program is still required to prepare and submit annual budget request documentation and congressional program status briefings, and the program office still reviews earned value management data and produces monthly program cost and status reports for the Air Force Program Executive Officer for Space. Additionally, the OSD officials stated that while at this time there were no plans to continue performing oversight analysis of the data required by the reports we cite, they agreed that finding a way to continue reporting these data would be valuable.

In addition, a new independent life-cycle cost estimate was not required for the program when it transitioned into the sustainment phase. As a result, OSD will not be able to rely on an updated life-cycle cost estimate for making long-term investment planning decisions. According to DOD officials, the latest life-cycle cost estimate for the program is not realistic, due in part to an optimistic launch vehicle production rate. Additionally, Air Force Space Command recently extended the life of the program by 10 years to fiscal year 2030, making the current life-cycle cost estimate— which only includes costs up until 2020—even less realistic. An updated independent life-cycle cost estimate for the program (developed once the ULA joint venture transition is completed and periodically updated thereafter to account for significant changes, such as fluctuations in launch vehicle demand) could help DOD to establish relevant and reliable baselines for the program as it moves forward and increase congressional insight into program costs. As we reported in 2006, costs for DOD space acquisitions over the past several decades have been consistently underestimated—sometimes by billions of dollars—because DOD typically does not update life-cycle cost estimates frequently enough to account for significant events and changes.[15] As we reported, at times the only mechanism that forced an updated estimate was DOD policy that the OSD Cost Analysis Improvement Group support the Nunn-McCurdy certification process for programs breaching a certain unit cost threshold. Prior to the August 2007 OSD decision to move EELV into the sustainment phase, the Air Force revised its acquisition strategy in part to obtain greater insight into contractor performance.

DOD Expanded Its Program Management Responsibilities

As part of its effort to increase its oversight and gain program knowledge, in 2005 the Under Secretary of the Air Force expanded the program office's management responsibilities when he approved a new acquisition strategy for the EELV program. Under the previous acquisition strategy, the government had less insight into contractor performance because it did not collect or have access to cost or pricing data and contractor performance data, making program costs difficult to estimate and limiting congressional insight into the program. Additionally, due to the size of the program office, there were limited opportunities to review and provide comments on launch vehicle technical design and production. The new acquisition strategy, "Buy 3," allows the government significantly more insight into the contractor's technical, cost, and schedule performance information because of the change in contract type and contracting strategy. For example, program officials told us that the launch capabilities contract now requires full earned value management[16] reviews that assess cost, schedule, and performance. In addition, contractor performance is more thoroughly evaluated and documented through assessment reports. Table 5 summarizes some of the additional primary program office responsibilities under the new acquisition strategy.

Table 5. Additional EELV Program Office Responsibilities Resulting from New Acquisition Strategy

Program office responsibilities	
Award fee criteria determination and evaluation	Sets criteria for award fees including fee determina-tion, evaluates contractor performance, and recomme-nds the amount of the award fee
Earned value management (EVM) reviews	Reviews new data requirements and cost data reports and complies with defense acquisition regulations for EVM data monitoring requirements
Report reviews/new reporting requirements	Reviews contractor cost data reports that include financial information on the costs that DOD contractors will incur while working on the program
	Generates a Contractor Performance Assessment Report that assesses and records a contractor's mance on a given contract during a specific Perforperiod; each assessment is based on objective facts and supp-orted by program and contract management data, such as cost performance reports, quality reviews, etc.
Integrated business reviews	Attends meetings to discuss cost performance and business issues resulting from EVM data reviews

Table 5. Continued

Program office responsibilities	
Contract negotia-tions (the procure-ng contracting off-icer, who resides in the program office, performs these activities)	On behalf of the government, conducts complex con-tract negotiations for both Atlas and Delta launch vehicles whenever a new launch is ordered and/or major changes are made to the infrastructure contracts
	Is responsible for all oversight operations at the launch ranges including operations and maintenance associated with facilities and facility upgrades; consequently, the program office must validate requirements for facility maintenance and upgrades and determine the roles and responsibilities for paying for these activities including:
	• determining cost-sharing responsibilities because facilities are used for both commercial and govern-mental purposes, and
	• addressing issues such as how to provide additional facilities for the contractor's use and making modifications to contractor property (leaky roofs, rusting tooling, etc.)
New entrant process	Certifies whether launch providers other than ULA meet program requirements, including key perfor-mance parameters and cost savings, via formal reviews of design, cost, and flight data
Engineering responsibilities	
Pedigree reviews	Reviews vehicle and component data packages to ens-ure that the subject articles have been manufactured and tested in accordance with approved processes
Design reviews	Conducts many design engineering analysis evalua-tions (as resources permit) due to earlier nonparticipa-tion in launch vehicle design, larger number of launch vehicle configurations than heritage launch vehicles, and executing first-time satellite-to-launch vehicle integration activities
Flight worthiness certification	Evaluates systems analyses, mission design and integration tasks, unplanned technical issue resolution, etc. to support certification of launch vehicle as ready to launch
Launch Operations	In conjunction with launch base personnel, conducts day-of-launch operation to execute mission assurance activities leading to launch and deployment of DOD payloads

ULA transition responsibilities	
Monitoring teams	Will monitor the technical transitions from both California to Colorado and from California and Colorado to Alabama
Senior leadership involvement	Are responsible for coordinating key transition goals such as the approval of the restructure proposal by the Under Secretary of Defense for Acquisition, Technology and Logistics, novation of contracts, and negotiation of leases

Source: GAO presentation of EELV program office data.

Staffing Shortage Impacts Program Office Ability to Manage the EELV Program

Although the increase in program management responsibilities has positioned the program office to better manage the EELV program, program officials told us that staffing in the EELV program office has not grown commensurately. As a result, they said the program now faces a shortage of government personnel with the right skills and experience to fulfill these new responsibilities. Program officials told us the staff shortage is a result of funding limitations, including the recent "force shaping" initiative to reduce Air Force military positions, as well as overall Air Force-wide recruiting challenges. According to program officials, this shortage of staff has hampered the program office's ability to complete the current workload. Currently, the program office has approximately 211 military and civilian personnel, as well as over 290 contractor staff.[17] However, given the change to the acquisition strategy, the program office has found it challenging to complete all tasks requiring a greater degree of specialized skills.

According to program officials, personnel shortages have occurred particularly in highly specialized areas such as avionics and launch vehicle groups. For example, program officials stated that 7 of 12 positions in the engineering branch for the Atlas group are now vacant. These engineers work on issues such as reviewing components responsible for navigation and control of the rocket. As table 6 shows, only half the government jobs in some key areas were projected to be filled.

Table 6. Summer 2008 Projected Staffing Levels within the EELV Program Office (as of March 2008)

Launch & Range Systems Wing staffing profile (as of March 2008)			
Program office group	**Authorized**	**Vacant**	**% Staffed**
Program Management	86	28	67.4%
Commander Wing	5	2	60.0
Management Operations	4	1	75.0
Program Control	39	16	59.0
Contracting	38	9	76.3
Engineering	26	13	50.0%
Atlas Group	32	16	50.0%
Engineering	12	7	41.7
Integration	11	7	36.4
Mission Support	7	2	71.4
Management	2	0	100.0
Delta Group	37	17	54.1%
Delta IV	12	4	66.7
Delta II	11	6	45.5
Delta Program Integration	13	7	46.2
Management	1	0	100.0
Space Lift Range Groups	129	26	79.8%
Spacelift Range Group (Los Angeles, AFB)	31	14	54.8
Spacelift Range Group-Eastern Range (Patrick AFB, FL)	31	2	93.5
Spacelift Range Group- Western Range (Vandenberg AFB, CA)	27	7	74.1
Range Sustainment (Peterson AFB, CO)	40	3	92.5
Total	**310**	**100**	**67.7%**

Source: GAO analysis of EELV program office data.

In addition to the perceived shortage of skilled personnel, turnover in leadership could also hamper the program office's ability to effectively manage the program. As GAO recently reported in congressional testimony on accountability issues in the acquisition environment, lack of program management tenure makes it difficult to hold program officials accountable for the business cases they are entrusted to manage and deliver.[18] According to the program office, the EELV program has recently experienced a loss in leadership, including the Wing Commander, or program director, responsible for the program office; the Delta Group Commander; the ULA Formation Program Manager; and the ULA Formation Contract Manager. While the Wing Commander has been replaced and the previously vacant Vice Commander position has been filled, there remains the loss of some of the most experienced personnel familiar with program execution and critical issues such as the ULA transition and current contract negotiations, according to the program office. Since the inception of the EELV program, the tenure for an EELV program director has ranged from 8 months to 49 months (the current program director has been with the program since June 2008), as shown in figure 3. With the exception of the first program director, this tenure falls short of DOD's workforce policy, which provides that the tenure period for program managers of major defense acquisition programs is the program milestone closest to 4 years or as tailored by the acquisition executive based on unique program requirements.[19]

While program officials could not specifically quantify the amount of work not being performed, they stated that the staffing shortage has affected the program's ability to complete programmatic and administrative tasks on time. According to program officials, the EELV program is also experiencing an increase in unplanned engineering work, which is considered critical to launch success and therefore prioritized over other tasks. These officials stated a portion of unplanned work has increased as a result of the lack of DOD's involvement in design engineering reviews early in the EELV program (during the design phase of the program under the previous acquisition strategy when DOD had limited oversight). For example, the program office became aware that the design assumptions for a Centaur inter-stage adapter[20] were flawed because they were based on old assumptions and no longer valid. As a result, the program office had to redesign the adapter, which currently does not meet load-carrying requirements. This redesign could have been averted had the Air Force been involved sooner, according to Air Force officials. In essence, the program has been trying to "catch up" by conducting engineering analyses or reviews that would have typically been conducted much earlier during the EELV design phase, which began in 1998 and ended in 2003. Additionally, program officials stated that an increase in the number of first-time integrations of new generations of satellites to EELVs adds to the number of design issues the program needs to address. Usually, the EELV program has been able to avoid delaying launches in order to complete unplanned work because the schedules for the programs developing the launch vehicle payloads have been delayed as a result of development difficulties. However, a launch delay resulting from the need to complete unplanned work has recently occurred. Specifically, an anomaly experienced during a June 2007 Atlas V launch resulted in a 2- month delay to the subsequent launch of another DOD satellite. Without enough personnel to complete the work, program officials are concerned that the EELV program may soon face additional launch delays.

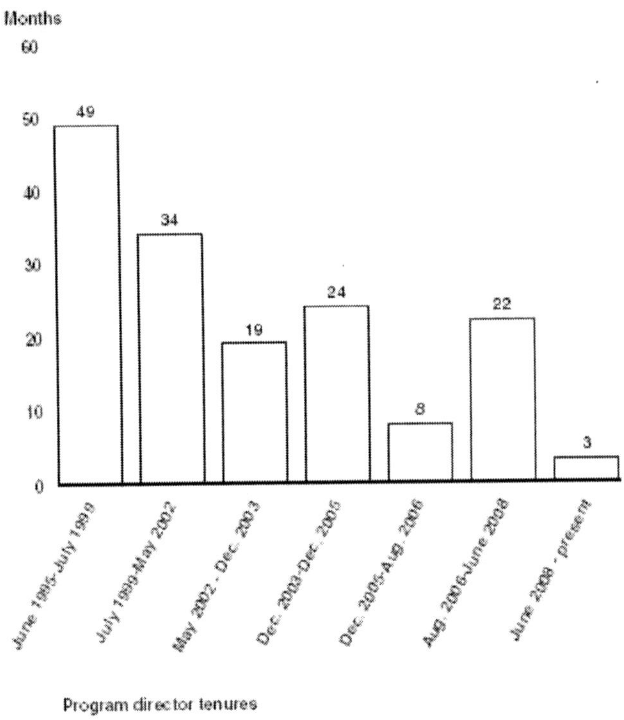

Source: GAO analysis of EELV program office data.

Figure 3. Tenures of EELV Program Directors from June 1995 to the Present

Air Force Initiatives Have Not Alleviated the EELV Program's Staffing Shortages

Air Force officials reported that recent Air Force personnel reductions had a disproportionate impact on the EELV program and its acquisition center. Beginning in 2006, the Air Force cut over 40,000 positions across the board in an effort to save money under Program Budget Decision (PBD) 720.[21] In addition, PBD 720 directed reductions to staffing levels of contractors that provide assistance to acquisition programs, particularly federally funded research and development centers (FFRDC) and systems engineering and technical assistance (SETA) contractors. Senior officials at the Manpower and Personnel Directorate at the Air Force Space and Missile Systems Center (SMC)—where EELV is managed—told us that PBD 720 affected SMC disproportionately because it relied heavily on FFRDC and SETA contractors to provide an array of services. According to SMC officials, PBD 720 may have further disproportionately affected the EELV program because it relied more heavily on uniformed personnel and assistance contractors than other space programs at SMC. EELV program officials told us they are in the process of awarding an engineering contract for technical assistance, but PBD 720 now limits the amount of contractor technical assistance the program can obtain due to funding limitations. Figure 4 shows how the program had increasingly relied on the Aerospace Corporation—its largest support contractor—for EELV engineering work over the past 10 years. This increase is largely due to efforts to prevent launch failures of legacy vehicles which occurred shortly after the EELV program was initiated. As figure 4 indicates, Aerospace Corporation's assistance to EELV, as shown by the number of personnel, increased slightly after 2006 but leveled off by 2008 as funding became more limited.[22]

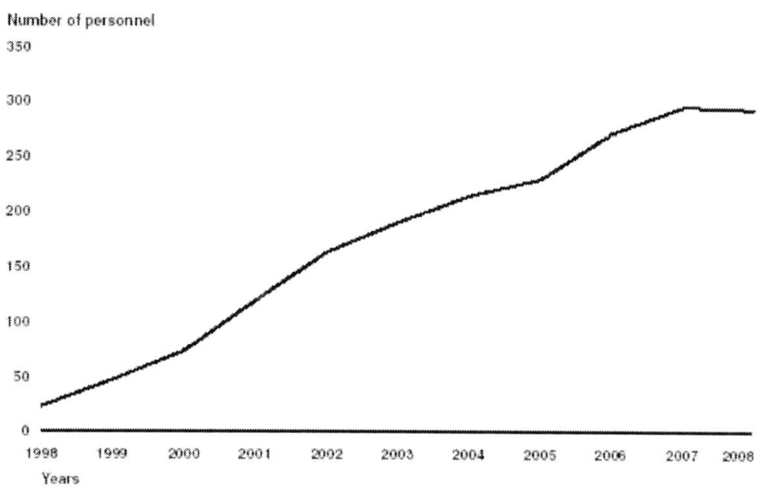

Source: GAO analysis of program office data.

Figure 4. Aerospace Corporation Assistance Provided to the EELV Program Office

Program officials cautioned that although Aerospace Corporation personnel have assisted with technical tasks the program office can not complete due to the staff shortage, the program can not rely on contractors to perform inherently governmental functions.[23] As we reported in September 2007, an increasing reliance on contractors to perform services for core government activities challenges the capacity of federal officials to supervise and evaluate the performance of these activities.[24] As one program official noted, mission assurance is an inherently governmental function, accountability for which should fall solely on the government.

Program officials stated that Air Force-wide efforts to address staffing shortages have not alleviated the EELV program's staffing shortages. Among these efforts was an attempt to counter the impacts of PBD 720 by converting some military positions to civilian positions—a move that would allow some of the military personnel who accepted the Air Force buy-out and left the service to return to the program as civilians. For example, when the Air Force initiated voluntary reductions by offering buy-outs to officers, 142 officers at SMC accepted. When SMC attempted to refill these positions as civilian positions, only 15 percent of the officers accepted the option to be rehired as civilians. In addition, the hiring process often takes too long to process such conversions, according to the program office. For example, when the EELV program was given eight military positions to convert to civilian positions, the program office lost all the candidates for these positions. In the time it took the program to create the new civilian positions, the candidates had found jobs elsewhere. (In a recent testimony, the Under Secretary of Defense for Acquisition, Technology and Logistics acknowledged that the hiring process takes too long and has contributed to the problem of retaining competent people in the acquisition field.) In addition, program officials said it was difficult to recruit and retain civilians in the Los Angeles, California, area given its high cost of living and its many opportunities for government contractor work for greater pay. Moreover, although manpower studies conducted by various DOD organizations have shown Air Force-wide staffing shortfalls, EELV program officials contend that these studies have not led to actions directed towards alleviating these shortfalls.

Table 7. Oversight Information Contained in Defense Acquisition Executive Summary Reports

Section	Description
Executive summary	Provides a summary-level assessment of and key information on the status of the program, including a synthesis of the issues that follow in the report (e.g., design problems exist that affect cost, schedule, and test and evalua-on). This section also provides information on general program status (e.g., acquisition category and date and type of next major program oversight review) and the program's cost, schedule, and performance baseline history.
Assessments	Represents the program manager's best independent judgment on program performance and serves an early indication of the risk in various aspects of the program. Using a color-coded rating system, the program manager applies performance ratings against the following indicators:
	• Performance characteristics: assesses the status of a broad range of mission performance criteria, including, but not limited to, essential physical, technical, operating, software, reliability, availability, maintainability, durability, manpower, training system effectiveness, and other similar characteristics needed to meet field or fleet needs.
	• Test and evaluation: assesses the status of system test planning, system testing, test article availability, test support, test center and range availability and funding, test success, and achievement of test schedules.
	• Logistics requirements and readiness objectives: assesses initiatives to achieve or maintain logistics management and support requirements (e.g., manpower, support equipment, training) and the ability of a system to undertake a specified set of missions or capabilities at planned peacetime and wartime utilization rates (e.g., for a missile system, established time to launch; for aircraft, previously agreed upon number of planes ready for take-off or time for take-off).
	• Cost performance: assesses the program's cost performance status based on performance to date, including the executability of the program within approved resources (based on cost and schedule performance status of the program's major contracts and the probable effects of those contracts on

Table 7. (Continued)

Section	Description
	cost estimates for future effort on the program); and considers potential unit cost reporting threshold breaches and status of cost reduction or cost/performance tradeoff initiatives.
	• Funding: assesses the overall adequacy and availability of programmed and budgeted funds by fiscal year, including effect of potential funding shortfalls, reductions or nonavailability due to congressional, Office of the Secretary of Defense, component, and/or cooperative allied country actions; program areas not funded to the approved acquisition program baseline; and whether the program is executable to the baseline, or if actual obligation rates are as planned.
	• Schedule performance: compares program's overall schedule performance and deliveries to date with program schedule milestones and annual delivery schedules; and considers the effect of schedule variations on major decision points, operational capability dates, and whether any major component of the system being developed or procured is not meeting the planned schedule.
	• Contracts: reviews all aspects of contract performance including technical and schedule achievement, cost performance, deliveries, contract change proposals and negotiations, and quality; identifies potential contract adjustments and their impact on ability to properly execute the contract; assesses all significant aspects of the contract award schedule, including definitization dates; and considers the effect of delays that threaten to extend major contract award dates that are on the critical path of program master schedule activities or that threaten to expose the government to unnecessary cost risk.
	• Production: assesses the overall status of the planning and execution of production and continuous process improvement activities, including all hardware and software aspects of the program; and considers key production requirements such as configuration management, technical data package availability, contractor capital investment, material availability, surge and mobilization planning, and capacity to meet delivery requirements.

Table 7. (Continued)

Section	Description
	• Management structure: assesses areas that do not fit the above assessment areas, such as the status of documentation, effect of problems from interrelated programs on this program, dependence of and problems for this program on government- or contractor-furnished equipment that are not managed or controlled by the program manager, adequacy of program office manpower to accomplish current or planned future requirements, relevant national security issues, or other areas of significance to the program office.
Program manager comments	Summarizes and explains observations, advisory comments, and potential or significant program problem areas for the categories shown in Defense Acquisition Executive Summary (DAES) Section 2, with emphasis on
	changes since the previous reporting period. At a minimum, this section is to include a description of problems and their significance relative to major program objectives; discussion of actions to be taken to accomplish program objectives and whether and how the program objectives need to be changed; an assessment of the status of corrective actions since the last DAES report; an assessment of any pending or proposed acquisition program baseline parameter changes, reasons for the changes, and associated risk assessments; and an assessment of changes made to any data parameters contained in approved program data (see DAES Section 5) that are not part of the acquisition program baseline.
Program executive officer/component acquisition executive[a] comments	Provides program executive office and component acquisition executive assessments and perspectives on the program, including changes in the relative level of risk associated with the program, the significance of problems reported by the program manager, the program manager's proposed corrective actions, the level of risk associated with these actions and other significant changes to the program; and comments on any pending or proposed acquisition program baseline parameter changes or changes to additional data elements that are not part of the acquisition program baseline but are contained in official program documentation and are integral to achieving the program objectives.

Table 7. (Continued)

Section	Description
Approved program data	Displays, in tabular form, key program parameters (including initial and current acquisition program baseline parameters as well as any additional data contained in other official program documents such as the acquisition plan or acquisition decision memoranda) in the following categories: performance characteristics, schedule milestones, and program acquisition cost. Data also include the program manager's current estimates for performance, schedule, and cost. This section is a key starting point for the program manager's evaluation of the program status in section 2.
	Performance characteristics: mission performance criteria needed to meet the significant objectives required by the end users (such as physical, technical, operational, software, survivability, reliability, and durability characteristics); initial and current acquisition performance baselines; results of demonstrated performance through testing; the program manager's current estimates relating to performance; and differences between the current estimates from major program objectives are to be explained in the appropriate part of Section 3.
	Schedule milestones: include initial and current approved schedule milestones as well as the program manager's current estimate for achieving schedule milestones. When the current estimates differ from the approved schedule milestones, the differences are explained in the appropriate part of Section 3.
	Program acquisition cost: displays total program costs for the entire acquisition process—from concept exploration and definition through production and deployment—as well as the program manager's current estimates of program costs. Costs include those related to research, development, test, and evaluation; procure-ment; program-specific military construction; and program acquisition-related operation and maintenance. This section also includes information on quantities of end items as well as unit costs.

Table 7. (Continued)

Program background data	Provides descriptive program-related information on total costs and total quantities for all years through the end of the acquisition phase for all DOD components, including information on and assessment of (as appropriate), program elements and procurement lines by appropriation, unit cost reporting data, procurement deliveries, con-tractor costs, international cooperative programs, other participating DOD components, and vulnerability assessments.
Supplemental contract cost information	Displays, in tabular form, summary-level contract identification, schedule, and performance information.
	Contract identification data: include contract name, number, and type; contractor name and location; program acquisition phase for which work is being done; negotiated contract cost or not-to-exceed value; target price; and ceiling price.
	Contract schedule data: include contract definitization date; work start date; dates of critical contract schedule milestones; and program manager's estimate of contract completion date.
	Contract performance data: include earned value management data for the most recent reporting period; the contractor's and program manager's current estimates of costs at completion; program manager comments on contract performance; and assessment of any changes to unit costs.
Funding summary	This section enables program offices to provide the approved funding profile and is intended to be used as the basis for identifying funding changes that could result in acquisition program baseline breaches, Nunn-McCurdy unit cost breaches, or other changes.

Source: GAO presentation of guidance on preparing Defense Acquisition Executive Summaries located on the Defense Acquisition University Web site.

[a] A program executive officer (PEO) is a military or civilian official who has responsibility for directing several major defense acquisition programs and for assigned major system and nonmajor system acquisition programs. A PEO has no other command or staff responsibilities within the component, and only reports to and receives guidance and direction from the DOD component acquisition executive (CAE). The CAE is a secretary of a military department or head of a military agency with the power of redelegation. In the Air Force, the delegated CAE is the Assistant Secretary of the Air Force (Acquisition). CAEs are responsible for all acquisition functions within their components.

CONCLUSIONS

Early in the EELV program, DOD was faced with two primary challenges: (1) reducing technical, design, and production risk to a point where it could avoid problems that caused previous, costly launch failures and (2) adopting an appropriate acquisition and supplier strategy that would help contain costs. By adding technical staff—primarily FFRDC—and strengthening quality assurance practices, DOD largely met the first challenge and has subsequently enjoyed a long string of successful launches though more experience is needed to fully prove reliability. At the same time, however, costs have increased significantly because assumptions supporting the first acquisition strategy never materialized and because DOD never pared down to one supplier as originally envisioned.

With the transition to ULA and the new acquisition strategy, it is hoped that cost savings can be achieved and the government can have more insight into cost drivers as well as technical risks. DOD has recognized that the transition itself presents enormous challenges and adds risk to the program by coordinating oversight efforts and expanding the program office's role in oversight. At the same time, however, OSD has limited these positive steps by designating EELV as being in sustainment, which effectively stops the formal transmittal of research and development and procurement data on program costs and vulnerabilities to entities involved in overseeing the program, including Congress. The Air Force may have also limited these steps by constraining program office staff. Given the broad span of changes affecting the program, the high cost of launching military and other government spacecraft, and assumptions that have been made about cost savings that will result from the consolidation of suppliers, DOD cannot afford to pare back acquisition and supplier oversight on any front.

End Notes

[1] The EELV program office's official name is the Launch and Range Systems Wing.

[2] The Nunn-McCurdy provision (10 U.S.C. § 2433) currently requires DOD to take specific actions when a major defense acquisition program's growth exceeds certain cost thresholds. Some of the key provisions of the law require, for example, that for major defense acquisition programs, (1) Congress must be notified when a program has an increase of at least 15 percent in program acquisition unit cost above the unit cost in the current baseline estimate and (2) the Secretary of Defense must certify the program to Congress when the program has unit cost increases of at least 25 percent of the current baseline estimate or at

least 50 percent of the original baseline estimate. The current law also includes cost growth thresholds from the program's original baseline estimate.

[3] Under Federal Acquisition Regulation (FAR) Part 12 procedures, the government had less insight into a contractor's costs since cost or pricing data cannot be required in the acquisition of commercial items.

[4] The FTC finalized the consent order on May 1, 2007, following a public comment period.

[5] NSSO facilitates the integration and coordination of defense, intelligence, civil, and commercial space activities.

[6] Of these 21 launches, 10 have been for DOD, 3 for civil government agencies (NASA and the National Oceanic and Atmospheric Administration), and 8 for commercial companies (based on payload).

[7] The fiscal year 2008 EELV program budget spans 6 years: fiscal years 2008 through 2013.

[8] The program office defines launch success as placing a satellite into orbit such that the payload is able to perform its mission. For example, during a June 2007 Atlas V launch of a National Reconnaissance Office (NRO) payload, an upper stage engine cut off 4 seconds early, placing the payload in a lower-than-desired orbit; however, the Air Force is confident the payload will be able perform its mission, and thus considers the launch a success.

[9] The Aerospace Corporation is a federally funded research and development center that provides objective technical analyses and assessments for military, civil, and commercial space programs.

[10] This 2002 report, entitled "Building Confidence in EELV," was an assessment of EELV launch risk and was prepared for the Air Force Assured Access Study.

[11] Booz Allen Hamilton, *Launch Mission Assurance Assessment Study* (April 2007). This study was requested by the Director, National Reconnaissance Office and the Commander, U.S. Air Force Space Command.

[12] 10 U.S.C. § 2325 (a) (1) (A) and (B).

[13] The Air Force portion of savings corresponds to the proportion of program costs for which it pays—70 percent (the NRO pays 30 percent)—and the resulting annual savings to the Air Force were projected at $105 million (70 percent of $150 million), beginning in 2011.

[14] Typically, for nonspace major defense acquisition programs, the sustainment phase begins when the acquired weapons or automated information systems have been fielded or deployed. In this phase, DOD oversight is normally reduced and program emphasis is on activities such as supply, maintenance, and transportation.

[15] GAO, *Space Acquisitions: DOD Needs to Take More Action to Address Unrealistic Initial Cost Estimates of Space Systems,* GAO-07-96 (Washington, D.C.: Nov. 17, 2006).

[16] Earned value management (EVM) is a program management tool that integrates the technical, cost, and schedule parameters of a contract. During the planning phase, an integrated baseline is developed by time-phasing budget resources for defined work. As work is performed and measured against the baseline, the corresponding budget value is "earned". Using this earned value metric, cost and schedule variances can be determined and analyzed. EVM provides significant benefits to both the government and the contractor. An EVM system is required on all DOD space program-related contracts meeting certain thresholds unless waived by the DOD Space Milestone Decision Authority. National Security Space Acquisition Policy, No. 03-01 (Dec. 27, 2004).

[17] The EELV program office was unable to provide us with the total number of personnel from all contractors providing assistance to the program office; therefore, we were only able to analyze numbers of Aerospace Corporation personnel. Military and civilian personnel figures are as of March 2008 and Aerospace personnel figures are as of May 2008.

[18] GAO, *Defense Acquisitions: Better Weapon Program Outcomes Require Discipline, Accountability, and Fundamental Changes in the Acquisition Environment*, GAO-08-782T (Washington, D.C.: June 3, 2008).

[19]Department of Defense Instruction No. 5000.66, *Operation of the Defense Acquisition, Technology and Logistics Workforce Education, Training, and Career Development Program* (Dec. 21, 2005), p.21.

[20] The inter-stage adapter connects the Atlas booster stage to the Centaur upper stage.

[21] Program Budget Decisions (PBD) reflect the decisions of the Office of the Secretary of Defense as to appropriate program and funding to be included in the annual defense budget request which, in turn, may be included in the President's Budget. PBD 720 outlines how manpower reductions will be accomplished to ensure the Air Force meets the force reduction mandated by PBD 720 and maintain mission capability. With PBD 720, the Air Force planned to eliminate 40,000 Active Duty, National Guard, and Reserve full-time equivalent (FTE) positions in order to self-finance the recapitalization and modernization of their aircraft, missile, and space inventories. PBD-720 also accelerated the retirement of a portion of the Air Force's aircraft inventory (F-117s, some B-52s, C-21s, and U-2s).

[22] In the late 1990s, the United States experienced a series of major, and costly, heritage launch vehicle failures. As a result, DOD commissioned a Space Launch Vehicles Broad Area Review to examine the causes of the launch vehicle failures and make recommendations to help ensure success for the remaining launches of heritage vehicles and for the transition to EELVs. Among the recommendations made by the review was for the Air Force to formulate a formal EELV launch risk management program, including an increased focus on launch success by identifying opportunities and resources needed for value-added government independent reviews of launch vehicle development and production. As a result of this recommendation, the EELV program office and Aerospace Corporation developed a rigorous launch mission assurance process, and Air Force investment in independent assessments increased substantially. For example, Aerospace Corporation involvement in the EELV program has increased from 25 personnel at EELV program initiation to the current level of about 290.

[23] The FAR provides that contracts must not be used for the performance of inherently governmental functions. FAR 7.503(a). Inherently governmental functions are so intimately related to the public interest that they should only be performed by government personnel. These functions include those activities which require either the exercise of discretion in applying Government authority or making value judgments in making decisions for the government, and should not be performed by contractors.

[24] GAO, *Defense Management: DOD Needs to Reexamine Its Extensive Reliance on Contractors and Continue to Improve Management and Oversight*, GAO-08-572T (Washington, D.C.: Mar. 11, 2008).

In: Future U.S. Space Launch Capabilities
Editors: Franz Lojdahl
ISBN: 978-1-60741-384-4

Chapter 3

SPACE LAUNCH VEHICLES: GOVERNMENT ACTIVITIES, COMMERCIAL COMPETITION, AND SATELLITE EXPORTS

Carl E. Behrens

ABSTRACT

Launching satellites into orbit, once the exclusive domain of the U.S. and Soviet governments, today is an industry in which companies in the United States, Europe, China, Russia, Ukraine, Japan, and India compete. In the United States, the National Aeronautics and Space Administration (NASA) owns and launches its space shuttle. Private sector companies provide launch services for other NASA launches and most Department of Defense (DOD) launches. Commercial customers purchase launch services from the U.S. companies or their competitors. Since the early 1980s, Congress and successive Administrations have taken actions, including passing several laws, to facilitate the U.S. commercial space launch services business. The Federal Aviation Administration (FAA) regulates the industry.

Forecasts in the 1990s suggesting significant increases in launch demand sparked plans to develop new launch v ehicles. NASA and DOD created government-industry partnerships to develop new reusable launch vehicles (RLVs) and "evolved" expendable launch vehicles (EELVs), respectively. (The space shuttle is the only RLV today. All other launch vehicles are "expendable" they can only be used once). Several U.S. private sector companies began developing their own launch vehicles.

Projections for launch services demand declined dramatically beginning in 1999, however. NASA's efforts to develop a new RLV to replace the shuttle faltered. DOD's new EELVs (Atlas V and Delta IV) began service, but, with reduced demand, the companies that build them (Lockheed Martin and Boeing) want more DOD funding to defray their costs. In 2005, the two companies announced plans to merge their EELV launch services for U.S. government customers. The joint venture, if approved by regulatory authorities, would be named the United Launch Alliance. Commercial launch services would not be affected.

Congress is debating the future of the space shuttle, which returned to flight in July 2005 after a two and one-half year hiatus following the 2003 *Columbia* tragedy. President Bush directed NASA to terminate the shuttle in 2010, but some want it to continue until a replacement is available. NASA also is embarking on a program to build two new "shuttle-derived" launch vehicles to implement the President's "Vision for Space Exploration" to return astronauts to the Moon and someday send them to Mars. The FY2007 request includes $837 million for the first of these new vehicles, the Crew Launch Vehicle.

In October 2004, Burt Rutan's SpaceShipOne suborbital spacecraft won the $10 million Ansari X-prize. Some believe this heralds an era of comparatively affordable space tourism. Congress passed a law in 2004 (P.L. 108-492) that creates a regulatory environment for space tourism. Concerns that China benefitted militarily from knowledge gained through commercial satellite launches in the 1990s led to changes in U.S. satellite export policy. The changes, especially returning control over such exports to the State Department from the Commerce Department, remain controversial because of what some claim is a negative impact on U.S. satellite manufacturing companies whose clients may choose European suppliers to avoid the U.S. export control regulations.

Most Recent Developments

The next launch of the space shuttle is now targeted for July 2006, according to a NASA announcement March 14, 2006. Shuttle flights were indefinitely postponed after the STS-114 mission in July-August 2005 because of a foam-shedding event from the shuttle's External Tank that was similar to what led to the loss of the space shuttle *Columbia* in 2003. Hurricane damage to the Michoud Assembly Facility (MAF) in New Orleans, LA, where the External Tanks are manufactured, slowed the work needed to resume flights. As directed by President Bush, NASA plans to retire the shuttle in 2010. It hopes to conduct 16 or 17 more shuttle launches by then. The FY2007 budget request for the shuttle is $4.1 billion, compared with $4.8 billion allocated to the shuttle in NASA's FY2006 Initial Operating Plan. Over the 5-year "runout" (FY2007-2011), the request significantly augments the space shuttle account because NASA concluded in 2005 that it needed $3-5 billion more for the shuttle than was included in the agency's FY2006 5-year spending plan. The FY2007 request redirects $2 billion from NASA's science programs and $1.5 billion from its exploration programs to fund the shuttle, as well as additional money for the International Space Station. NASA's Stennis Space Center, MS, where the space shuttle's main engines are tested, also suffered hurricane damage. NASA estimates that it needs $760 million to cope with damage from Hurricane Katrina. The Administration requested $325 million, and Congress provided $350 million, for NASA for hurricane relief in the DOD appropriations and hurricane recovery act (P.L. 109-148). NASA is hoping that it will be able to request the remaining needed funds in a supplemental.

NASA's FY2007 request also includes $837 million to build a new launch vehicle, the Crew Launch Vehicle (CLV), to support President Bush's Vision for Space Exploration. The CLV is a "shuttle-derived" design. Furthermore, NASA is allocating $500 million through FY2009 to assist commercial companies develop and demonstrate launch vehicles that could service the International Space Station.

For FY2007, DOD is requesting $937 million for procurement of EELVs and $19 million for research and development (R and D).

BACKGROUND AND ANALYSIS

U.S. Launch Vehicle Policy

The National Aeronautics and Space Administration (NASA), the Department of Defense (DOD), and the private sector have developed expendable launch vehicles (ELVs) — they can only be used once — to place satellites in orbit. NASA also developed the partially reusable space shuttle. U.S. ELVs currently in use include Atlas (manufactured by Lockheed Martin), Delta (Boeing), and Pegasus and Taurus (Orbital Sciences Corporation). Delta IV and Atlas V are the most recent additions to the fleet, and were developed through DOD's Evolved Expendable Launch Vehicle (EELV) program.

From "Shuttle-Only" to "Mixed Fleet"

In 1972, President Nixon approved NASA's plan to build the first reusable launch vehicle, called the space shuttle, and directed that it become the nation's primary launch vehicle, replacing all the ELVs except Scout (later discontinued for unrelated reasons). This would have made NASA and DOD dependent on a single launch vehicle, but the resulting high launch rate was expected to reduce the cost per flight significantly. The shuttle was first launched in 1981, and was declared operational in 1982. The phase-out of the ELVs began, but in 1984 the Air Force successfully argued that it needed a "complementary" ELV as a backup to the shuttle for "assured access to space" and initiated what is now known as the Titan IV program. Production lines for the Delta and Atlas began to close down, and it was expected that only the shuttle, Scouts, and Titan IVs would be in use by the mid-1980s.

Everything changed on January 28, 1986, however, when the space shuttle *Challenger* exploded 73 seconds after launch. Apart from the human tragedy, the *Challenger* accident deeply affected U.S. space launch policy, demonstrating the vulnerability of relying too heavily on a single system. Many military and civilian satellites had been designed to be launched on the shuttle, and could not have been transferred to ELVs even if the ELVs were not already being phased out. The remaining ELVs had their own problems in 1986. A Titan exploded in April and a Delta failed in May, which also grounded Atlas because of design similarities. Consequently, the Reagan Administration revised U.S. launch policy from primary dependence on the shuttle to a "mixed fleet" approach where a wide variety of launch vehicles are available. The shuttle is used principally for missions that require crew interaction, while ELVs are used for launching spacecraft. President Reagan also decided that commercial payloads could not be flown on the shuttle unless they were "shuttle-unique" (capable of being launched only by the shuttle or requiring crew interaction) or if there were foreign policy considerations. That action facilitated the emergence of a U.S. commercial space launch industry whose participants had long argued that they could not compete against government-subsidized shuttle launch prices. The White House and Congress had taken steps beginning in 1983 to assist in developing a commercial space launch services business, including President Reagan's 1983 designation of the Department of Transportation as the agency responsible for facilitating and regulating the commercial space launch sector. Passage of the 1984 Commercial Space Launch Act (P.L. 98-575), the Commercial Space Launch Act Amendments of 1988 (P.L. 100-657), and the Commercial Space Act of

1998 (P.L. 105-303) also have helped. But removing the shuttle as a competitor is seen as the major factor in fostering the U.S. commercial space launch business.

CLINTON ADMINISTRATION POLICY

On August 5, 1994, President Clinton released a National Space Transportation Policy that gave DOD lead responsibility for improving ELVs and NASA lead responsibility for upgrading the space shuttle and technology development of new reusable launch vehicles. The policy also set guidelines for the use of foreign launch systems, the use of excess ballistic missile assets for space launch, and encourages an expanded private sector role in space transportation R and D.

GEORGE W. BUSH ADMINISTRATION POLICY

On December 21, 2004, President Bush authorized a new U.S. Space Transportation Policy that supersedes the 1994 Clinton policy. A fact sheet on the Bush policy was released on January 6, 2005 [http://www.ostp.gov/html/SpaceTransFactSheetJan2005.pdf]. The new policy calls both for continued government support for space transportation capabilities, and for capitalizing on the U.S. private sector's "entrepreneurial spirit." DOD is directed to maintain the capability to develop, evolve, operate and purchase services for space transportation systems, infrastructure, and support activities necessary for national security requirements. NASA is directed to maintain the same capability for the civil sector, and to engage in development activities only for those requirements that cannot be met by capabilities being used by the national security or commercial sectors. The policy also directs NASA, in cooperation with DOD, to develop options to implement the President's January 2004 Vision for Space Exploration. NASA subsequently decided to build shuttle-derived launch vehicles. The policy states that the government will continue to support both EELVs (Boeing's Delta IV and Lockheed Martin's Atlas V) until the Secretary of Defense certifies to the President that U.S. assured access to space can be maintained without both providers.

U.S. LAUNCH VEHICLE PROGRAMS AND ISSUES

NASA's Space Shuttle Program

The Space Transportation System (STS) — the space shuttle — is a partially reusable launch vehicle and is the sole U.S. means for launching humans into orbit. It consists of an airplane-like Orbiter, with two Solid Rocket Boosters (SRBs) on each side, and a large, cylindrical External Tank (ET) that carries fuel for the Orbiter's main engines. The Orbiters and SRBs are reused; the ET is not. NASA has three remaining spaceflight-worthy Orbiters: *Discovery*, *Atlantis*, and *Endeavour*.

The Challenger and Columbia Tragedies

A total of 114 shuttle launches have taken place since April 1981. Two ended in tragedy, each killing seven astronauts. In 1986, the space shuttle *Challenger* exploded 73 seconds after launch because of the failure of a seal (an O-ring) between two segments of an SRB. In 2003, the space shuttle *Columbia* disintegrated as it returned to Earth after 16 days in orbit (see CRS Report RS21408). *Columbia* broke apart from aerodynamic forces after the left wing was deformed from the heat of gases that entered the wing through a hole caused during launch by a piece of foam insulation that detached from the ET. The *Columbia* Accident Investigation Board (CAIB) found that the tragedy was caused by technical and organizational failures, and made 29 recommendations, 15 of which it said should be completed before the shuttle returned to flight (see CRS Report RS21606 for a synopsis of the CAIB report).

Sean O'Keefe, NASA's Administrator from December 2001-February 2005, said NASA would comply with the CAIB recommendations. He established an RTF Task Group, chaired by two former astronauts, Tom Stafford and Dick Covey, to oversee NASA's implementation of the CAIB's 15 RTF recommendations. The Stafford/Covey Task Group [http://www.returntoflight.org] did not address management and culture changes, and was not tasked to determine whether the shuttle was ready to return to flight. Its assignment only was to evaluate NASA's compliance with the CAIB recommendations for RTF. The Task Group ultimately concluded that NASA met the intent of 12 of the 15 CAIB RTF recommendations, but not the other three: eliminating debris shedding from the External Tank, hardening the Orbiter so it can better survive debris impacts, and developing an on-orbit method of repairing the shuttle's thermal protection system. Dr. Michael

Griffin, who became NASA Administrator in April 2005, said that NASA and contractor personnel are those responsible and accountable for determining if and when the shuttle is ready for RTF, and would not commit to meeting every CAIB recommendation.

Return to Flight (RTF)

NASA launched the space shuttle *Discovery* on the first of two "Return to Flight" (RTF) missions — STS-114 — on July 26, 2005 and it successfully landed on August 9. On July 27, however, NASA announced that a piece of foam had detached from STS-114's ET during launch, similar to what happened to *Columbia*. Cameras and other sensors on *Discovery* and on the International Space Station — to which *Discovery* was docked for much of its mission — imaged the Orbiter and determined that it was not damaged, but further shuttle launches have been suspended until the problem is resolved. Meanwhile, the images revealed that two "gapfillers" — ceramic coated fabric placed between thermal protection tiles — were protruding on the belly of the Orbiter that could have affected aerodynamic heating during reentry. One of the *Discovery* astronauts removed them during a space walk. The second RTF mission — STS-121 — was scheduled for September 2005, but now planned for July 2006.

The United Space Alliance (USA)

In 1995, NASA decided to turn most shuttle operations over to a "single prime contractor" — the United Space Alliance (USA), a limited liability company owned 50-50 by Boeing and Lockheed Martin. USA was created to pull together the 86 separate contracts with 56 different companies under which the shuttle program was then operating. NASA signed a $7 billion, six-year Space Flight Operations Contract (SFOC) with USA on September 26, 1996, with the goal of reducing shuttle operational costs while ensuring safety. The SFOC contract has been extended to 2006. NASA officials assert that SFOC has saved NASA $1 billion a year compared to what the costs would have been without it. Contracts for the External Tank, Solid Rocket Boosters (SRBs), and Space Shuttle Main Engines (SSMEs) have not been incorporated into SFOC. NASA manages those contracts: with Lockheed Martin for the External Tank; ATK Thiokol for the SRBs, and Pratt and Whitney Rocketdyne (owned by United Technologies) for the SSMEs.

The Shuttle's Future

As discussed below, NASA attempted unsuccessfully for many years to develop a "2nd generation" reusable launch vehicle (RLV) to replace the shuttle (which is the 1st generation RLV). In a November 2002 amendment to its FY2003 budget request, NASA announced a new space transportation strategy that indicated the shuttle would continue flying until at least 2015, and perhaps 2020 or beyond. The *Columbia* tragedy, and President Bush's January 2004 announcement of a new "Vision for Space Exploration" — to return astronauts to the Moon by 2020 and someday send them to Mars — forced NASA to revise that plan.

The President's Vision calls for the shuttle program, which absorbs approximately 25% of NASA's annual budget, to be terminated in 2010. A primary motivation is to make that funding available to implement other aspects of the Vision, although there also is concern about shuttle safety. Congress has been debating the Vision, including its impact on the shuttle and on U.S. human access to space. Some Members want to terminate the shuttle earlier than 2010 because they feel it is too risky and/or that the funds should be spent on accelerating the Vision. Others want to retain the shuttle at least until a new spacecraft, the Crew Exploration Vehicle (CEV), is available to take astronauts to and from the ISS. Under the plan announced by the President, CEV would not be ready at least until 2014, leaving a multi-year gap during which U.S. astronauts would have to rely on Russia for access to the ISS. NASA Administrator Griffin hopes to accelerate CEV development and have it ready by 2012, but has made clear that he views 2010 as a firm deadline for terminating the shuttle.

At the beginning of 2005, NASA officials indicated 28 shuttle flights were needed to complete ISS construction. When he became Administrator, Dr. Griffin stated that such a number was not achievable. In the fall of 2005, NASA announced a new plan showing 18 shuttle missions to the ISS, and possibly one more to service the Hubble Space Telescope (see CRS Report RS21767). As part of its FY2007 budget request, NASA reduced to 16 the number of shuttle flights it now plans to the ISS (plus one possible flight to Hubble). The two that were cut were logistics flights taking cargo to the ISS. NASA is hope that commercial launch service provider will develop vehicles that can perform that task. The original versions (H.R. 3070/S. 1281) of what became the 2005 NASA authorization act (P.L. 109-155) had conflicting language about the future of the shuttle. The original Senate bill directed NASA not to terminate the shuttle until a replacement was available; the House version directed NASA not to fly the shuttle after December 31, 2010.

The final law states that it is U.S. policy to have human access to space on a continuous basis, and directs NASA to submit several related reports to Congress.

A related issue is what steps NASA and USA must take to make certain that the shuttle workforce retains the needed personnel and skills to ensure the shuttle flies safely through its remaining years of service. A 2005 GAO report (GAO-05-230) concluded that NASA needs to better position itself to address future shuttle workforce needs.

Shuttle Budget

NASA's FY2006 Initial Operating Plan shows that NASA plans to spend $4.8 billion on the shuttle in FY2006. NASA Administrator Griffin conceded in 2005 that NASA's FY2006 budget request (prepared prior to his arrival at NASA) contained estimated ("placeholder") figures for shuttle funding in FY2008-2010 that were significantly lower than what is actually needed, and the program was underfunded by $3-5 billion over those three years. The FY2007 budget request adds $2 billion to the shuttle program to help compensate for that shortfall. For FY2007, NASA is requesting $4.1 billion for the shuttle program. The projected figures for FY2008-2010 are $4.1 billion, $3.8 billion, and $3.7 billion respectively, with a FY2011 projected funding level of $147 million as the program is terminated.

"Shuttle-Derived" Launch Vehicles

On September 19, 2005, NASA announced its implementation plan for the Vision for Space Exploration. Under the plan, NASA will develop two new launch vehicles that are derived from the space shuttle system. One is for launching the new Crew Exploration Vehicle (CEV); the other for launching large amounts of cargo. The first is designated the Crew Launch Vehicle (CLV), and would use a single SRB, augmented by a new "upper stage," with the CEV on top. It is informally referred to as the "single stick" design. The cargo launch vehicle, referred to as a "heavy lift" vehicle because it will be able to launch 125 tons of cargo to Earth orbit, would use a modified ET, supplemented by SRBs (like the shuttle), with the cargo spacecraft on top. NASA Administrator Griffin and Dr. Ronald Sega, Under Secretary of the Air Force, agreed that NASA would develop these two shuttle-derived vehicles for implementing the Vision, but use DOD's Evolved Expendable Launch Vehicles (EELVs) for its other spacecraft in the 5-20 metric ton range to the maximum extent possible (although new commercially-developed launch vehicles may compete with the EELVs if they become available).

NASA's Efforts to Develop New Reusable Launch Vehicles (RLVs)

U.S. expendable and reusable launch systems remain expensive and less efficient and reliable than desired. DOD and NASA initiated several efforts in the late 1980s and early 1990s to develop new systems, but each was terminated in turn because Congress or the agencies themselves were not convinced that the required investment had sufficient priority. In response to the 1994 Clinton policy, DOD's Evolved Expendable Launch Vehicle (EELV) program, and NASA's Reusable Launch Vehicle (RLV) program, were initiated.

Proponents believe that RLV technology can dramatically lower the cost of accessing space. NASA's efforts to develop a "2nd generation" RLV to replace the shuttle have not fared well, however. Starting in 1995, NASA pursued two "X" (for "experimental") flight test programs: X-33 and X-34. X-33 was a joint program with Lockheed Martin to build a subscale prototype of a large RLV based on single-stage-to-orbit (SSTO) technology. The SSTO concept involves a rocket that can attain orbit with only one stage (instead of two or more as is common today) carrying people or cargo. X-34 was a small RLV "testbed" to demonstrate reusable two-stage-to-orbit technologies, which was being built under a traditional contract with Orbital Sciences Corporation. NASA terminated X-33 and X-34 in March 2001 because the cost to complete them was too high relative to the benefits. NASA spent $1.2 billion on X-33, and Lockheed Martin said that it spent $356 million of its own funding. NASA spent $205 million on X-34.

NASA restructured its RLV program in 2000 (as part of its FY2001 budget request) and initiated the Space Launch Initiative (SLI). It then restructured the SLI program in 2002, and terminated it following President Bush's announcement of the Vision in January 2004. The goal was to develop RLV technology that would be "10 times safer and crew survivability 100 times greater, all at one-tenth the cost of today's space launch systems." The failure of the X-33 and X-34 programs, and of the National AeroSpace Plane (NASP) program before them, made some observers skeptical about NASA's ability to develop a 2nd generation RLV. In documentation accompanying a November 2002 budget amendment, NASA conceded that a new RLV lacked economic justification. SLI was restructured into two components: building an Orbital Space Plane (OSP), a spacecraft (not a launch vehicle) to take crews to and from the space station, and developing "Next Generation Launch

Technology" (NGLT). Concurrent with President Bush's announcement of the Vision, NASA terminated SLI.

DOD's Evolved Expendable Launch Vehicle (EELV) Program

DOD began what is now known as the EELV program in FY1995 (P.L. 103-335) with a $30 million appropriation. EELV was first formally identified in DOD's FY1996 budget. Two EELVs were developed in joint government-private sector programs: Boeing's Delta IV and Lockheed Martin's Atlas V. Both vehicles have successfully entered service (although the first launch of the Delta IV "heavy" — i.e., the version that can launch the greatest amount of mass — in December 2004 did not reach its intended orbit). The goal of the EELV program is to reduce launch costs by 25%.

In 1996, the Air Force selected Lockheed Martin and McDonnell Douglas (later bought by Boeing) for development contracts worth $60 million. Originally, one of those companies would have been selected in 1998 to develop the EELV. In November 1997, responding to indicators at the time that the commercial space launch market would be larger than expected, DOD announced that it would help fund development of both Atlas V and Delta IV. In October 1998, DOD awarded Boeing $1.88 billion for the Delta IV ($500 million for further development plus $1.38 billion for 19 launches), and awarded Lockheed Martin $1.15 billion for the Atlas V ($500 million for further development plus $650 million for 9 launches). The companies were expected to pay the rest of the development costs themselves. (Boeing officials state that Boeing invested $2.5 billion in design, development, and infrastructure for the Delta IV, of which the company wrote off $2 billion.)

In 2000, however, new market forecasts showed a reduction in expected commercial demand, and DOD reevaluated its EELV strategy. It renegotiated the contracts with both companies, including relieving Lockheed Martin (reportedly at the company's request) of the requirement to build a launch pad at Vandenberg AFB, CA. Each company built a launch pad for its vehicle at Cape Canaveral, FL for east coast launches. Both were expected to build them at Vandenberg for west coast launches (which launch site to use is determined by the type of orbit required by the satellite), but under this agreement, only Boeing would be able to launch from the west coast, giving it a monopoly on those EELV contracts. The companies also approached DOD to obtain additional government funding because of the downturn in the commercial market. This is called "assured access to space" in the sense of assuring that both companies remain in the EELV business so DOD has redundancy in capability should one of the launch vehicles experience difficulties. The FY2004 DOD authorization act (P.L. 108-136) codified "assured access" as U.S. policy.

Following revelations of ethics violations by Boeing in 2003, including some related to the EELV program, DOD suspended three Boeing business units from eligibility for new government contracts, shifted seven existing launch contracts from Boeing to Lockheed Martin, and disqualified Boeing from bidding for three new launch contracts. Boeing withdrew the Delta IV from competition for commercial contracts because it did not believe it could successfully compete. DOD reinstated the plan for an Atlas V launch pad at Vandenberg, so Boeing no longer would have a monopoly on west coast launches. The Boeing suspension was lifted on March 4, 2005.

The EELV program breached the "Nunn-McCurdy" limit of 25% cost growth, which requires DOD to cancel or restructure the program, or certify that it is essential to national security. In April 2004, DOD made that certification. In its report on the FY2005 DOD appropriations bill (H.Rept. 108-553), the House Appropriations Committee directed DOD to study whether both families of EELVs were really needed (H.Rept. 108-553). The December 2004 Bush space transportation policy directs DOD to continue to support both EELVs until the Secretary of Defense certifies to the President that assured access can be maintained without two EELV providers.

On May 2, 2005, Boeing and Lockheed Martin announced they planned to merge the production, engineering, test, and launch operations associated with EELV launch services for the U. S. government. If approved by U.S. and foreign regulatory bodies, both vehicles would be built at a single location (Boeing's facility in Decatur, AL), while engineering and administrative activities would be consolidated at Lockheed Martin's offices in Denver, CO. The joint venture, to be owned 50-50 by the two companies, would be called United Launch Alliance. The companies said the merger would save $100-150 million per year for the U.S. government. The merger does not affect commercial launch services.

For FY2007, DOD is requesting $937 million for procurement of EELVs and $19 million for research and development (R and D).

Private Sector Launch Vehicles (Including Space Tourism and the X-Prize)

Several entrepreneurial U.S. companies have been attempting to develop RLVs through private financing. Many have encountered difficulties in obtaining financing from the financial markets, and some have sought government loan guarantees or tax credits. Some have received limited direct government funding through various contracts. One company, SpaceX, headed by Elon Musk (creator of PayPal), asserts that it will dramatically reduce the cost of reaching orbit with its partially reusable Falcon launch vehicle. The first Falcon launch was scheduled for 2004, but has been delayed by a variety of factors.

Some companies are focusing on suborbital rockets instead of those that can attain orbit, anticipating that suborbital space tourism will be a substantial market. Twenty seven teams from seven countries competed in the "Ansari X-Prize" contest [http://www.xprize.com] to win $10 million by becoming the first privately-financed company to launch a vehicle capable of carrying three people (one person actually had to be aboard) to an altitude of 100 kilometers (62.5 miles), return safely to Earth, and repeat it within two weeks using the same vehicle. On October 4, 2004, Burt Rutan's Scaled Composites Inc. won the X-prize with the SpaceShipOne vehicle, financed by Microsoft co-founder Paul Allen. Sir Richard Branson, head of the Virgin Group, is licensing the SpaceShipOne technology. He founded a company, Virgin Galactic, to offer commercial suborbital flights, and someday orbital flights. He reportedly expects to invest about $100 million in the new spaceships and associated ground infrastructure, and charge $200,000 per person per flight.

P.L. 108-492 creates a regulatory environment for commercial human space flight ("space tourism"). It sets requirements for protecting third parties, and for the crews of commercial spacecraft. There are few regulations for passengers ("spaceflight participants"), based on the philosophy that anyone who is willing to take the risk to fly on these new spacecraft should be allowed to do so as long as they are informed of the vehicle's safety record. If there are a significant number of accidents, or incidents that could have led to accidents, the FAA may set further passenger regulations. After eight years, the FAA may set any regulations it wishes. The FAA published a proposed rule in the December 29, 2005 *Federal Register* seeking public comment. Others believe that commercial human space flight should be regulated more strongly. H.R. 656 would strengthen safety requirements for passengers. A House Science subcommittee held a hearing on April 20, 2005, where Mr. Rutan sharply criticized FAA's regulatory process.

U.S. Commercial Launch Services Industry

Congressional Interest

Congress has been debating issues involving the domestic launch services industry for many years. Part of the debate has been focused on satellite export issues (discussed below). Another part concerns what the government should do to stimulate development of new launch vehicles by the private sector. That debate focuses on whether tax incentives or loan guarantees should be created for companies attempting to develop lower cost launch vehicles. Tax incentive advocates argue that loan guarantee programs allow the government to pick winners and losers; loan guarantee advocates argue that tax incentives are insufficient to promote necessary investment in capital intensive projects. Congress created (Title IX, FY2003 DOD appropriations Act, P.L. 107-248) a loan guarantee program for companies developing commercial, reusable, in-orbit space transportation system, but such systems are not launch vehicles (they move satellites from one orbit to another) and are not discussed further here. In 2004, Congress passed a law (P.L. 108-428) extending third-party liability indemnification for the commercial space launch industry for five more years. The government indemnifies commercial space launch companies for third-party claims between $500 million and $2 billion. Other 2004 legislation relating to regulation of space tourism is discussed above. In the 109th Congress, H.R. 3643 would make spaceports, like airports, eligible for tax exempt bonds.

Foreign Launch Competition

Europe, China, Russia, Ukraine, India, and Japan offer commercial launch services in competition with U.S. companies. Most satellites are manufactured by U.S. companies or include U.S. components and hence require export licenses, giving the United States considerable influence over how other countries participate in the commercial launch services market. The United States negotiated bilateral trade agreements with China, Russia, and Ukraine on "rules of the road" for participating in the market to ensure they did not offer unfair competition because of their non-market economies. Launch quotas were set in each of the agreements. However, President Clinton terminated the quotas for Russia and Ukraine in 2000, and the agreement with China expired at the end of 2001.

Europe

The European Space Agency (ESA) developed the Ariane family of launch vehicles. The first test launch of an Ariane was in 1979; operational launches began in 1982. ESA continued to develop new variants of Ariane. Ariane 5 is the only version now in use. ESA also is developing a smaller launch vehicle, Vega, whose first launch is expected in 2005. Operational launches are conducted by the French company Arianespace. Arianespace conducts its launches from Kourou, French Guiana, on the northern coast of South America. Arianespace also markets Russia's Soyuz launch vehicle and ESA is planning to build a launch site for Soyuz at Kourou. In 1985, a U.S. company (Transpace Carriers Inc.) filed an unfair trade practices complaint against Arianespace, asserting that European governments were unfairly subsidizing Ariane. The Office of the U.S. Trade Representative (USTR) investigated and found that Europe was not behaving differently from the United States in pricing commercial launch services (then offered primarily on the government-owned space shuttle). The incident raised questions about what "rules of the road" to follow in pricing launch services. In the fall of 1990, USTR and Europe began talks to establish such rules of the road and assess how to respond to the entry of non-market economies into the launch services business. The only formal negotiating session was held in February 1991.

Each side is concerned about how much the respective governments subsidize commercial launch operations, but another controversial topic (not formally part of the talks) was whether Arianespace should be able to bid for launches of U.S. government satellites, which now must be launched on U.S. launch vehicles as a matter of U.S. policy. Arianespace wants that restriction lifted. France and other European governments do not have written policies requiring the use of Ariane for their government satellites. However, the member governments of ESA originally agreed to pay a surcharge of as much as 15-20% if they chose Ariane. The surcharge led some cost-conscious European governments to buy launch services from other (notably U.S.) suppliers. In the fall of 1995, ESA's member governments reached agreement with Arianespace to reduce the surcharge to encourage use of Ariane. ESA itself gives preference to using Ariane, but is not legally constrained from using other launch vehicles. Arianespace has encountered significant financial difficulties both because of the constrained market, and because of the failure of a new, more capable variant of the Ariane 5 in 2002. In May 2003, the ESA Council of Ministers adopted a European Guaranteed Access to Space (EGAS) program, providing 960 million euros for Arianespace to fix that variant of Ariane 5 (it successfully returned to flight in February 2005), and acquire Ariane 5 launch vehicles through 2009, while the commercial launch market is down.

China

The People's Republic of China offers several versions of its Long March launch vehicles commercially. China poses special issues not only because of its non-market economy, but because of technology transfer and political concerns. Launch services are offered through China Great Wall Industry Corp. (CGWIC). Because the United States currently will not issue export licenses for satellites or satellite components destined for China (see below), the Chinese commercial space launch program is sharply constrained.

U.S.-China Bilateral Trade Agreements for Launch Services

In 1989, China and the United States signed a six-year bilateral trade agreement restricting the number of Chinese commercial space launches to ensure China, with its nonmarket economy, did not unfairly compete with U.S. companies. A new seven-year agreement was reached in 1995, and amended in 1997. The agreement expired on December 31, 2001. While the agreements were in force, they established quotas on how many commercial satellites China could launch each year, and included pricing provisions to try to ensure that China did not unfairly compete with U.S. commercial launch service providers because of its non-market economy.

U.S. Satellite Exports to China: 1988-1997

In September 1988, the U.S. government agreed to grant three export licenses for satellites manufactured by Hughes to be launched by CGWIC. The Reagan Administration granted the licenses on the conditions that China sign three international treaties related to liability for satellite launches and other subjects; agree to price its launch services "on a par" with Western companies; and establish a government-to-government level regime for protecting technology from possible misuse or diversion. China met the conditions and the two countries signed a six-year agreement in January 1989. The now-defunct Coordinating Committee on Multilateral Export Controls (COCOM) approved the licenses that March.

On June 5, 1989, after the Tiananmen Square uprising, President George H. W. Bush suspended all military exports to China. At the time, exports of communications satellites were governed by the State Department's Munitions List. The satellites counted as military exports and the licenses were suspended. Then Congress passed language in the FY1990 Commerce, Justice, State and Judiciary appropriations (P.L. 101-162) and the 1990-91 Foreign Relations Authorization Act (P.L. 101-246, Section 902) prohibiting the export of U.S.-built satellites to China unless the President reported to Congress that (1) China had achieved certain political and human rights reforms, or (2) it was in the national interest of the United States. In December 1989, President Bush notified Congress that export of the satellites was in the national interest and the licenses were reinstated. The satellites were launched by China in 1990-1992.

A different issue arose in 1990. China signed a contract to launch an Arabsat Consortium satellite for $25 million, much less than what many considered "on a par" with Western companies. The main competitor, Arianespace, turned to both the French and U.S. governments to prohibit export of the satellite. No formal action was taken by the United States. In 1991, the Arabsat Consortium terminated the contract with the Chinese and signed an agreement with Arianespace, so the case became moot, but the issue of what constituted "on a par" remained. China argued that because its costs are so low, it could offer lower prices and still adhere to international norms as to what costs are included in setting the price. Yet another issue arose in 1991 — linkage of satellite export licenses with U.S. concern over China's ballistic missile proliferation policies. On April 30, 1991, the Bush Administration approved final export licenses for two satellites and for U.S. components of another, but to emphasize its concern about Chinese missile proliferation, disapproved export of U.S. components for a communications satellite China itself was building. On June 16, 1991, the White House announced that it would not approve any further export licenses for commercial satellite launches. On July 17, the State Department identified CGWIC as one of two Chinese entities engaged in missile technology proliferation activities that require the imposition of trade sanctions under the Arms Export Control Act, including denial of license applications for export items covered by the Missile Technology Control Regime (MTCR). Although the MTCR does not cover satellites (only satellite launch vehicles, which are close cousins of ballistic missiles), the identification of CGWIC as a cause of concern complicated

China's marketing plans. China agreed to adhere to the MTCR, and the sanctions were lifted on February 21, 1992. In May 1992, INTELSAT agreed to launch at least one satellite on a Chinese launch vehicle. On September 11, 1992, the State Department notified Congress that it was waiving legislative restrictions on U.S. exports for six satellite projects with China. Many observers saw the move as a conciliatory gesture in the wake of the U.S. decision to sell F-16s to Taiwan.

On August 25, 1993, however, the U.S. government again imposed sanctions against China for ballistic missile proliferation activities, and the State Department said that satellite exports would not be permitted. The State Department announced October 4, 1994, it would lift the sanctions after China pledged to abide by the MTCR. During this period, tensions were acute between those viewing the sanctions as harmful to U.S. business interests and those seeking to prevent sensitive technology from reaching China and/or to punish China for MTCR infractions. The debate centered on whether the satellites should be governed by the State Department (Munitions List) or the Commerce Department (Commerce Control List). Some responsibility for export of commercial communications satellites was transferred from the State Department to the Commerce Department in 1992; in October 1996 primary responsibility was transferred to Commerce.

In January 1995, the launch of the Hughes-built APStar-2 satellite failed in-flight. Falling debris killed 6 and injured 23 on the ground. On February 6, 1996, President Clinton approved the export of four satellites to China for launch, despite concerns about China exporting nuclear weapons-related equipment to Pakistan. On February 14, 1996, a Long March 3B rocket carrying the INTELSAT 708 communications satellite built by Loral malfunctioned seconds after liftoff, impacting the ground and spreading debris and toxic fumes over the launch site and a nearby village. The Chinese reported 6 dead and 57 injured, but other reports suggested a higher figure. Some customers, including INTELSAT, canceled contracts. In May 1997, USTR stated that it believed China violated the pricing provisions of the bilateral agreement for the launching of Agila 2 for the Philippines. Chinese officials disagreed. On September 10, 1997, the *Washington Times* published a story that Chinese and Russian entities (including CGWIC) were selling missile technology to Iran. China denied the allegations.

Satellite Exports to China: 1998-2000 (Including the "Loral/Hughes" Issue, the Cox Committee Report, and Lockheed Martin)

On February 18, 1998, the President notified Congress that it was in the national interest to export Loral's Chinasat 8 satellite to China. On April 4, 1998, the *New York Times* reported that a 1997 classified DOD report alleged that Space Systems/Loral (part of Loral Space and Communications) and Hughes Electronics' satellite manufacturing division (then a subsidiary of General Motors; now Boeing Satellite Systems) provided technical information to China that improved the reliability of Chinese nuclear missiles. The assistance was provided in the wake of the February 1996 INTELSAT 708 launch failure (see above). The INTELSAT satellite was built by Loral, which participated in an inquiry into the accident at the request of insurance companies seeking assurances that the Chinese had correctly diagnosed and solved the cause of the failure. Loral formed a review committee that included representatives of other satellite companies, including Hughes. According to Loral, the review committee did not itself investigate the accident, but listened to Chinese officials explain their investigation and then wrote a report. Loral conceded that a copy of the report was given to the Chinese before it was provided to the State Department, in violation of Loral's internal policies. Loral says it notified the State Department when it learned that the Chinese had been given a copy. According to media sources, DOD's 1997 report says that the companies provided technical information in violation of Loral's export license. The companies insist they did not violate the licenses. The Justice Department investigated, and expanded the probe to include Hughes' response to the 1995 APStar-2 failure. A grand jury reportedly was empaneled in 1999. The government reached a civil settlement with Loral on January 9, 2002 where Loral agreed to pay a $14 million civil fine, and spend $6 million on strengthening its export compliance program. On December 26, 2002, the State Department charged Hughes Electronics and Boeing Satellite Systems with 123 export violations. The companies settled with the government on March 5, 2003, accepting a civil penalty of $20 million in cash, and $12 million in credits for money already spent ($4 million), or that will be spent ($8 million), on export program enhancements.

Many hearings on the "Loral/Hughes" issue were held by various House and Senate committees. In addition, the House established the Select Committee on U.S. National Security and Military/Commercial Concerns with the People's Republic of China, chaired by Representative Cox. The Cox committee concluded that Hughes and Loral deliberately transferred technical information and know-how to China during the course of accident investigations. The committee investigated other cases of China acquiring other U.S. technical information and made 38 recommendations.

The FY2000 DOD authorization act (P.L. 106-65) included language implementing many of the Cox committee recommendations. In brief, the Department of Justice must notify appropriate congressional committees when it is investigating alleged export violations in connection with commercial satellites or items on the munitions list if the violation is likely to cause significant harm or damage to national security with exceptions to protect national security or ongoing criminal investigations; companies must be provided with timely notice of the status of their export applications; enhanced participation by the intelligence community in export decisions is required; adequate resources must be provided for the offices at DOD and the State Department that approve export licenses; individuals providing security at overseas launch sites do not have to be DOD employees, but must report to a DOD launch monitor; and DOD must promulgate regulations concerning the qualifications and training for DOD space launch monitors and take other actions regarding those monitors and the records they maintain.

In February 1999, the Clinton Administration denied Hughes permission to export two satellites to China for launch. Export permission for those satellites (called APMTs) had been granted in 1997, but Hughes changed the spacecraft design, necessitating new export approval. That application was denied. On May 10, 2000, the White House made its first certification to Congress under the new process detailed in the FY1999 DOD authorization bill, approving the export to China of satellite fuels and separation systems for the Iridium program. On August 18, 2000, the State Department stated it would continue the suspension of a technical assistance agreement for Loral regarding launch of the Chinasat 8 satellite because the concerns that initiated the suspension in December 1998 had not been rectified. In January 2001, *Space News* reported that the Chinasat 8 export application was returned to Loral without action.

In April 2000, it became known that Lockheed Martin also was under investigation, in this case for performing a technical assessment, without an export license, of a Chinese "kick motor" used to place a satellite into its final orbit. On June 14, 2000, the State Department announced it had reached agreement with Lockheed Martin involving $13 million in penalties — $8 million that the company will pay over a four-year period and $5 million that was suspended and that the company can draw upon to fund a series of remedial compliance measures specified in the consent agreement.

Satellite Exports to China: 2001-Present

In July 2001, Senators Helms, Thompson, Shelby, and Kyl wrote to President Bush reportedly asking the President not to grant waivers for the export of satellites to China. As noted earlier, such waivers are required under the FY1990-91 Foreign Relations Authorization Act (P.L. 101-246). At the time, attention was focused on two European companies (Astrium and Alenia Spazio) that had built satellites for two multinational satellite organizations (INTELSAT and EUTELSAT, respectively) that were scheduled for launch by China. The satellites contain U.S. components. In August 2001, INTELSAT canceled its contract with Astrium for the APR-3 satellite, citing several factors, including the delay in obtaining U.S. export approval. EUTELSAT switched the launch of its satellite to Europe's Ariane. Other satellites being manufactured by U.S. companies, however, such as Chinasat 8 and another being built by Loral (Apstar-5, for APT Satellite Co.), or containing U.S. components may require waivers in the future (see CRS Report 98-485 for a list of pending satellite exports). The FY2002 Commerce, Justice, State Appropriations Act (P.L. 107-77), the FY2003, FY2004, and FY2005 Consolidated Appropriations Acts (P.L. 108-7, 108-199, and 108-447), and the FY2006 Science, State, Justice, Commerce Appropriations Act (P.L. 109-108) require 15 days notice to Congress before processing licenses for exporting satellites to China.

Russia

U.S. policy prohibited U.S.-built satellites from being exported to the Soviet Union. Following the collapse of the Soviet Union, President George H. W. Bush said he would not oppose Russia launching an International Maritime Satellite Organization (Inmarsat) satellite and the United States would negotiate with Russia over "rules of the road" for future commercial launches. Discussions in the fall of 1992 led to agreement in principle in May 1993; the agreement was signed on September 2, 1993, after Russia agreed to abide by the terms of the MTCR (see below). On January 30, 1996, the countries amended the agreement. Prior to Russia's first launch of a U.S.-built satellite, a Technology Safeguard Agreement among the United States, Russia, and Kazakstan (where the launch site is located) was signed in January 1999. A similar agreement for launches from Russia's Plesetsk, Svobodny, and Kapustin Yar launch sites was signed in January 2000.

The 1993 agreement was signed only after Russia agreed to comply with the MTCR in a case involving a Russian company, Glavkosmos, that planned to sell rocket engine technology to the Indian Space Research Organization

(ISRO). The United States declared it violated the MTCR and imposed two-year sanctions against Glavkosmos and ISRO. In June 1993, the United States threatened to impose sanctions against Russian companies that did business with Glavkosmos. The two countries finally agreed that Russia would cease transferring rocket engine technology (the engines themselves were not at issue) to India. As noted, on September 10, 1997, the *Washington Times* published a story that Russian and Chinese entities, including the Russian Space Agency, were selling missile technology to Iran. In July 1998, Russia publicly identified nine entities that might be engaged in illegal export activities. The United States imposed sanctions against seven of them on July 28 and three more on January 12, 1999. The State Department said the United States would not increase the quota on geostationary launches that Russia could conduct under the 1996 agreement unless Russian entities ceased cooperating with Iran's ballistic missile program. The launches are conducted primarily by a U.S.-Russian joint venture composed of Lockheed Martin and Russia's Khrunichev and Energia, companies that were not among those sanctioned. Lockheed Martin was anxious to have the quota raised to 20 and eventually eliminated. On July 13, 1999, the White House raised the quota to 20, and eliminated it on December 1, 2000. (*Wall Street Journal,* December 1, 2000, p. A4).

Ukraine

Ukraine offers commercial launch services, chiefly as part of the Sea Launch joint venture among Boeing, Ukraine's Yuzhnoye, Russia's Energomash, and Norway's Kvaerner. The Sea Launch vehicle consists of a Ukranian two-stage Zenit rocket with a Russian third stage. The vehicle is launched from a mobile ocean oil rig built by Kvaerner. The rig is stationed in Long Beach, CA, where the launch vehicle and spacecraft are mated, and then towed into the ocean where the launch takes place. The United States and Ukraine signed a bilateral trade agreement in February 1996, that would have expired in 2001, but President Clinton terminated it on June 6, 2000, in recognition of "Ukraine's steadfast commitment to international nonproliferation norms." The first successful commercial launch was in October 1999. In 1998, Boeing agreed to pay $10 million for not abiding by export regulations in its dealings with Russia and Ukraine. Sea Launch announced plans in October 2003 to offer launches from Baikonur using Zenit beginning in 2005; the effort is called Land Launch.

India

India conducted its first successful orbital space launch in 1980. Its ASLV and PSLV launch vehicles can place relatively small satellites in low Earth orbit. India is developing a larger vehicle (GSLV) capable of reaching geostationary orbit. The GSLV, which uses Russian cryogenic engines that were the subject of a dispute between the United States and Russia (discussed earlier), made it first operational flight in September 2004.

Japan

Japan successfully conducted the first launch of its H-2 launch vehicle in 1994, the first all-Japanese rocket capable of putting satellites in geostationary orbit. Previous rockets used for this purpose were based on U.S. technology and a 1969 U.S.-Japan agreement prohibited Japan from launching for third parties without U.S. consent. With the H-2, Japan was freed from that constraint. H-2 was not cost effective, and encountered technical problems that led the Japanese government to abandon it in 1999. A new version, H2A, successfully completed its first launch in August 2001. In 2002, the Japanese government announced that it would privatize production of the H2A Mitsubishi Heavy Industries took over development and marketing. H-2A launches are conducted from Tanegashima, on an island south of Tokyo. In June 1997, the Japanese government reached agreement with the fishing industry to allow more launches from Tanegashima. Fishermen must evacuate the area near the launch site during launches. The agreement extends from 90 to 190 the number of days per year that launches may be conducted, and permits up to eight launches a year instead of two.

Satellite Exports: Agency Jurisdiction and Other Issues

Between 1992 and 1996, the George H. W. Bush and Clinton Administrations transferred responsibility for decisions regarding export of commercial satellites from the State Department to the Commerce Department. A January 1997 GAO report (GAO/NSIAD-97-24) examines that decision. In response to concerns about the launch of satellites by China (discussed above), Congress directed in the FY1999 DOD authorization bill (P.L. 105-261) that export control responsibility be returned to the State Department effective March 15, 1999. Which agency should control these exports remains controversial.

The 108th Congress debated, but did not clear, legislation on this topic. The House International Relations Committee (HIRC) reported H.R. 1950 (H.Rept. 108-105, Pt. 1), the FY2004 State Department Authorization Act, with language that would have left the decision on agency jurisdiction to the President if the export was to a NATO country or major non-NATO ally. Exports to China would have remained under State Department jurisdiction. The House Armed Services Committee rejected the HIRC language in its markup of the bill (H.Rept. 108-105, Pt. 3), however, and the House-passed version did not include that language. There was no further legislative action on that bill.

Some of the controversy reflects concerns of the aerospace and space insurance industries that the regulations are being implemented too broadly and vigorously. DOD officials and others have cited potential harm to the U.S. defense industrial base if U.S. exports are stifled, too. One concern is the length of time needed to obtain State Department approval. Section 309 of the FY2000 State Department authorization act (incorporated into the FY2000 Consolidated Appropriations Act, P.L. 106-113) directed the Secretary of State to establish an export regime with expedited approval for exports to NATO allies and major non-NATO allies. The rules took effect July 1, 2000. In May 2000, the State Department reportedly notified France that it would not apply strict technology export control on satellites to be launched by Ariane (*Space News*, May 29, 2000, p. 1). The Security Assistance Act (P.L. 106-280) reduced from 30 days to15 days the time Congress has to review decisions on exporting commercial communications satellites to Russia, Ukraine, and Kazakhstan, making the time period the same as for NATO allies.

The Satellite Industry Association (SIA) released figures in May 2001 showing U.S. satellite manufacturers losing market share to foreign companies. SIA and others attributed that loss in part to the shift in jurisdiction to State, which they assert creates uncertainty for satellite customers over when and whether export licenses will be approved. The trade publication *Space News* reports on the number of new commercial satellite orders awarded world-wide each year. According to that source, U.S. companies won 19 of the 22 contracts in 2001; three of the four in 2002; nine of the 16 in 2003; and nine of the 12 in 2004.

LEGISLATION

109th Congress, 1st Session

P.L. 109-108, H.R. 2862. FY2006 Science, State, Justice, Commerce appropriations bill (includes NASA). Reported from House Appropriations Committee June 10, 2005 (H.Rept. 109-118); passed House June 16. Reported from Senate Appropriations Committee June 23 (S.Rept. 109-88); passed Senate September 15. Conference report (H.Rept. 109-272) passed House November 9, Senate November 16. Signed into law November 22, 2005.

P.L. 109-148, H.R. 2863. FY2006 DOD appropriations bill. Reported from House Appropriations Committee June 10, 2005 (H.Rept. 109-119); passed House June 20. Reported from Senate Appropriations Committee September 29 (S.Rept. 109-141); passed Senate October 7. Conference report (H.Rept. 109-359) passed House December 19, Senate December 21 (amended). Signed into law December 30, 2005.

P.L. 109-155, S. 1281. FY2007-2008 NASA authorization bill. H.R. 3070 would have authorized NASA funding for FY2006-2007; reported by the House Science Committee July 18 (H.Rept. 109-173), passed House July 22, 2005. S. 1281, as originally passed, authorized NASA funding for FY2006-2010; reported by Senate Commerce Committee July 26 (S.Rept. 109-108), passed Senate September 28, 2005. Conference report (H.Rept. 109-354), which authorizes funding for FY2007-2008, passed House December 17, Senate December 21, 2005. Signed into law December 30, 2005.

P.L. 109-163, H.R. 1815. FY2006 DOD authorization bill. Reported from House Armed Services Committee May 20, 2005, H.Rept. 109-89; passed House May 25. S. 1042 reported from Senate Armed Services Committee May 17 (S.Rept. 109-69); passed Senate November 15. Conference report (H.Rept. 109-360) passed House December 19, Senate December 21. Signed into law January 6, 2006.

In: Future U.S. Space Launch Capabilities
Editors: Franz Lojdahl
ISBN: 978-1-60741-384-4

Chapter 4

SPACE STATIONS

Marcia S. Smith

ABSTRACT

Congress continues to debate NASA's International Space Station (ISS), a permanently occupied facility in Earth orbit where astronauts live and conduct research. Congress appropriated approximately $35 billion for the program from FY1985-2005. The initial FY2006 ISS request was $2.180 billion: $1.857 billion for construction and operations and $324 million for research to be conducted by ISS crews. In a July budget amendment, NASA transferred $168 million for ISS Crew/Cargo Services to another part of the NASA budget and reduced the ISS request commensurately. The final version of the FY2006 appropriations bill that includes NASA (H.R. 2862) cuts $80 million from the originally submitted budget, and NASA now plans to spend $306 million, instead of $324 million, on ISS research in FY2006.

The space station is being assembled in Earth orbit. ISS segments, crews, and cargo are taken into orbit by Russian or U.S. spacecraft. ISS has been permanently occupied by successive "Expedition" crews rotating on 4-6 month shifts since November 2000. "Expedition 12" is now aboard. Cost growth and schedule delays have characterized the program since its inception. The grounding of the space shuttle fleet after the 2003 *Columbia* tragedy and the July 2005 *Discovery* "Return to Flight" mission is further affecting schedule, operations, and cost. Most of the remaining ISS segments are designed to be launched by the shuttle and construction therefore is suspended. President Bush's January 2004 "Vision for Space Exploration" also is affecting the ISS program. He directed that the shuttle program be terminated in 2010, and changed the focus of ISS-based research to only that which supports his "Moon/Mars" Vision instead of the broadlybased program that was planned.

Canada, Japan, and several European countries became partners with NASA in building the space station in 1988; Russia joined in 1993. Except for money paid to Russia, there is no exchange of funds among the partners. Europe, Canada, and Japan collectively expect to spend about $11 billion of their own money. A reliable figure for Russian expenditures is not available.

In 1993, when the current space station design was adopted, NASA said it would cost $17.4 billion for construction (not including launch or other costs). That estimate grew to $24.1-$26.4 billion, leading Congress to legislate spending caps on part of the program in 2000. The estimate then grew by almost another $5 billion, leading NASA (at White House direction) to cancel or indefinitely defer some hardware to stay within the cap. NASA exceeded the cap in FY2005, however.

Controversial since the program began in 1984, the space station has been repeatedly redesigned and rescheduled, often for costgrowth reasons. Congress has been concerned about the space station for that and other reasons. Twenty-two attempts to terminate the program in NASA funding bills were defeated, however (3 in the 106th Congress, 4 in the 105th Congress, 5 in the 104th , 5 in the 103rd , and 5 in the 102nd). Three other attempts in broader legislation in the 103rd Congress also failed.

Current congressional debate focuses on the impact of space shuttle-related delays, and the future of ISS in light of President Bush's new exploration initiative.

MOST RECENT DEVELOPMENTS

The "Expedition 12" crew, American John McArthur and Russian Valery Tokarev, continue their work aboard the International Space Station (ISS). They arrived via the Russian Soyuz TMA-7 spacecraft on October 3. They replaced the Expedition 11 crew, which returned to Earth in the Soyuz TMA-6 spacecraft on October 10. ISS crews are taken to and from the ISS on Russian Soyuz spacecraft while launches of the U.S. space shuttle are grounded. Shuttle launches have been postponed until at least May 2006 because of a problem during the most recent launch, STS-114, in July 2005 (see CRS Report RS21408). Until then, U.S. astronauts will remain dependent on Russia for access to the ISS. Russia has been providing crew transport and "crew return" (i.e., a "lifeboat" capability for emergencies) services to NASA at no cost under a 1996 agreement. Russia's obligations under that agreement have been fulfilled with the launch of Soyuz TMA-7. It will remain docked with the ISS during the Expedition 12 mission, scheduled to end in April 2006. In the future, NASA will have to pay for Soyuz services, but has been prohibited from doing so unless Russia complies with the Iran Nonproliferation Act (INA). The House and Senate have passed a bill (S. 1713) that would amend the INA to allow NASA to pay for such services through January 1, 2012 (see CRS Report RS22270).

The original FY2006 request for the ISS was $2.180 billion: $1.857 billion for construction and operations (including $160 million for ISS Crew/Cargo Services) and $324 million for research. In the final version of the FY2006 Science, State, Justice and Commerce (SSJC) appropriations bill (H.R. 2862), which includes NASA, Congress cut $80 million from the ISS program, including $60 million from ISS Crew/Cargo Services. The House passed a FY2006-2007 NASA authorization bill (H.R. 3070, H.Rept. 109-173); it does not specify an amount for the ISS program. The Senate passed a FY2006-2010 NASA authorization bill (S. 1281, S.Rept. 109-108) that adds $100 million for FY2006 and calls for enhancing the use of ISS for research. On July 15, NASA amended its FY2006 request, shifting the funds for ISS Crew/Cargo services (which it now identifies as $168 million) into a different budget account, and commensurately reducing the ISS request. The conference report on H.R 2862 does not reflect the budget amendment. See CRS Report RL32988 for details on the FY2006 NASA budget request.

BACKGROUND AND ANALYSIS

Introduction

NASA launched its first space station, Skylab, in 1973. Three crews were sent to live and work there in 1973-74. It remained in orbit, unoccupied, until it reentered Earth's atmosphere in July 1979, disintegrating over Australia and the Indian Ocean. Skylab was never intended to be permanently occupied. The goal of a permanently occupied space station with crews rotating on a regular basis was high on NASA's list for the post-Apollo years. In 1969, Vice President Agnew's Space Task Group recommended a permanent space station and a reusable space transportation system (the space shuttle) to service it as the core of NASA's program in the 1970s and 1980s. Budget constraints forced NASA to choose to build the space shuttle first. When NASA declared the shuttle "operational" in 1982, it was ready to initiate the space station program.

In his January 25, 1984 State of the Union address, President Reagan directed NASA to develop a permanently occupied space station within a decade, and to invite other countries to join. On July 20, 1989, the 20th anniversary of the first Apollo landing on the Moon, President George H. W. Bush gave a major space policy address in which he voiced his support for the space station as the cornerstone of a long-range civilian space program eventually leading to bases on the Moon and Mars. That "Moon/Mars" program, the Space Exploration Initiative, was not greeted with enthusiasm in Congress, primarily due to budget concerns, and ended in FY1993, although the space station program continued.

President Clinton dramatically changed the character of the space station program in 1993 by adding Russia as a partner to this already international endeavor. That decision made the space station part of the U.S. foreign policy agenda to encourage Russia to abide by agreements to stop the proliferation of ballistic missile technology, and to support Russia economically and politically as it transitioned from the Soviet era. The Clinton Administration strongly supported the space station within certain budget limits.

President George W. Bush, prompted in part by the February 2003 space shuttle *Columbia* tragedy, made a major space policy address on January 14, 2004, directing NASA to focus its activities on returning humans to the Moon and someday sending them to Mars. Included in this "Vision for Space Exploration" is a plan to retire the space shuttle in 2010. The President said the United States would fulfill its commitments to its space station partners, but the details of how to accomplish that without the shuttle were not announced.

THE SPACE STATION PROGRAM: 1984-1993

NASA began the current program to build a space station in 1984 (FY1985). In 1988, the space station was named *Freedom*. Following a major redesign in 1993, NASA announced that the *Freedom* program had ended and a new program begun, though NASA asserts that 75% of the design of the "new" station is from *Freedom*. The new program is simply referred to as the International Space Station (ISS). Individual ISS modules have various names. (Some refer to the facility as "Space Station Alpha," but that is not its formal name). ISS is a laboratory in space for conducting experiments in near-zero gravity ("microgravity"). A broadly based research program was planned, but

President Bush's Vision for Space Exploration would limit U.S. research on ISS to that which is needed to support the goal of sending humans back to the Moon and to Mars. From FY1985- FY2005, Congress appropriated approximately $35 billion for the space station program.

Space Station *Freedom*

When NASA began the space station program in 1984, it said the program would cost $8 billion (FY1984 dollars) for research and development (R and D — essentially the cost for building the station without launch costs) through completion of assembly. From FY1985-1993, Congress appropriated $11.4 billion to NASA for the *Freedom* program. Most of the funding went for designing and redesigning the station over those years. Little hardware was built and none was launched. Several major redesigns were made. A 1991 redesign evoked concerns about the amount of science that could be conducted on the scaled-down space station. Both the White House Office of Science and Technology Policy (OSTP) and the Space Studies Board (SSB) of the National Research Council concluded that materials science research could not justify building the space station, and questioned how much life sciences research could be supported. They criticized the lack of firm plans for flying a centrifuge, considered essential to this research. NASA subsequently agreed to launch a centrifuge, but, in September 2005, decided that it was not needed (see below).

Cost estimates for *Freedom* varied widely depending on when they were made and what was included. *Freedom* was designed to be operated for 30 years. As the program ended in 1993, NASA's estimate was $90 billion (current dollars): $30 billion through the end of construction, plus $60 billion to operate it for 30 years. The General Accounting Office (GAO) estimated the total cost at $118 billion, including 30 years of operations.

In 1988, after three years of negotiations, Japan, Canada and nine European countries under the aegis of the European Space Agency (ESA) agreed to be partners in the space station program. A government-to-government Intergovernmental Agreement (IGA) was signed in September, and Memoranda of Understanding (MOUs) between NASA and its counterpart agencies were signed then or in 1989. The partners agreed to provide hardware for the space station at their own expense, a total of $8 billion at the time.

1993 Redesign — the Clinton Administration Restructuring

In early 1993, as President Clinton took office, NASA revealed $1 billion in cost growth on the *Freedom* program. The President gave NASA 90 days to develop a new, less costly, design with a reduced operational period of 10 years. A new design, *Alpha,* emerged on September 7, 1993, which NASA estimated would cost $19.4 billion. It would have used some hardware bought from Russia, but Russia was not envisioned as a partner. Five days earlier, however, the White House announced it had reached preliminary agreement with Russia to build a joint space station. Now called the International Space Station (ISS), it superseded the September 7 *Alpha* design. NASA asserted it would be a more capable space station and be ready sooner at less cost to the United States. Compared with the September 7 *Alpha* design, ISS was to be completed one year earlier, have 25% more usable volume, 42.5 kilowatts more electrical power, and accommodate six instead of four crew members.

In 1993, President Clinton pledged to request $10.5 billion ($2.1 billion a year) for FY1994-1998. NASA said the new station would cost $17.4 billion to build, not including money already expended on the *Freedom* program. That estimate was derived from the $19.4 billion estimate for the September 7 *Alpha* design minus $2 billion that NASA said would be saved by having Russia in the program. The $2.1 billion and $17.4 billion figures became known as "caps," though they were not set in law. (See Cost Caps below).

THE INTERNATIONAL SPACE STATION (ISS): 1993-PRESENT

The International Space Station program thus began in 1993, with Russia joining the United States, Europe, Japan, and Canada. The 1993 and subsequent agreements with Russia established three phases of space station cooperation and the payment to Russia of $400 million, which grew to $473 million. (NASA transferred about $800 million to Russia for space station cooperation through this and other contracts.)

During Phase I (1995-1998), seven U.S. astronauts remained on Russia's space station *Mir* for long duration (several month) missions with Russian cosmonauts, Russian cosmonauts flew on the U.S. space shuttle seven times, and nine space shuttle missions docked with *Mir* to exchange crews and deliver supplies. Repeated system failures and two life-threatening emergencies on *Mir* in 1997 raised questions about whether NASA should leave more astronauts on *Mir*, but NASA decided *Mir* was sufficiently safe to continue the program. (As discussed below, *Mir* was deorbited in 2001.) Phases II and III involve construction of the International Space Station itself, and blend into each other. Phase II began in 1998 and was completed in July 2001; Phase III is underway.

ISS Design, Cost, Schedule, and Lifetime

ISS is being built by a partnership among the United States, Russia, Europe, Japan, and Canada. The 1988 Intergovernmental Agreement was renegotiated after Russia joined the program. The new version was signed in 1998. The IGA is a treaty in all the countries except the United States, where it is an Executive Agreement. The IGA is implemented through Memoranda of Understanding (MOUs) between NASA and its counterpart agencies. Brazil is not a partner in ISS, but agreed to participate through a bilateral agreement with NASA. Boeing is the U.S. prime contractor.

NASA originally stated that ISS would be operated for 10 years after assembly was completed, with a possibility for 5 additional years if the research was considered worthwhile. Using the original schedule, assembly would have been completed in 2002, with operations at least through 2012. The completion of assembly slipped to 2006, but President Bush restructured the space station program in 2001, and it was not clear when assembly would be "completed." NASA briefing charts in March 2003 showed space station operations possibly continuing until 2022. Under President Bush's January 2004 "Vision for Space Exploration," however, NASA plans to complete its utilization of ISS in 2016 (though the other partners may continue to use it after that time).

ISS segments are launched into space on U.S. or Russian launch vehicles and assembled in orbit. The space station is composed of a multitude of modules, solar arrays to generate electricity, remote manipulator systems, and other elements that are too numerous to describe here. Details can be found at [http://spaceflight.nasa.gov/home/index.html]. Six major modules are now in orbit. The first two were launched in 1998: Zarya ("Sunrise," a Russian-built, U.S.-owned, module with guidance, navigation, and control systems) and Unity (a U.S. "node" connecting other modules). Next was Zvezda ("Star," a Russian module that serves as the crew's living quarters) in 2000. Destiny (a U.S. laboratory), Quest (a U.S. airlock), and Pirs ("Pier," a Russian docking compartment) arrived in 2001. Among the other modules awaiting launch are laboratory modules built by Russia, Europe, and Japan, and two more "nodes" built by Europe. (Zarya counts as a U.S. module because NASA paid Russia to build it. Some of the European- and Japanese-built hardware counts as U.S. because they are built under barter agreements where Europe and Japan produce hardware NASA needs instead of paying cash to NASA for launch and other ISS-related services.) Ordinarily, the U.S. space shuttle takes crews and cargo back and forth to ISS. The shuttle system is currently grounded because of problems that occurred during the July 2005 launch of STS-114 (see CRS Report RS21408). Russian Soyuz spacecraft are also used to take crews to and from ISS, and Russian Progress spacecraft deliver cargo, but cannot return anything to Earth (Progress is not designed to survive reentry into the Earth's atmosphere). A Soyuz is always attached to the station as a lifeboat in case of an emergency.

The schedule for launching segments and crews is called the "assembly sequence" and has been revised many times. At the end of the Clinton Administration, the assembly sequence showed completion of assembly ("assembly complete") in April 2006. The most recent assembly sequence was released after a January 2005 "Heads of Agency" meeting in Montreal, but it does not include launch dates, only the order in which the launches will go. It does list "Establishment of a Permanent Crew of Six (January 2009)," followed by nine shuttle launches to assembly complete. Under the Vision, ISS construction is to be completed by 2010, but NASA Administrator Griffin has indicated that a sufficient number of shuttle flights may not be able to be launched in that time period. He intends to terminate the shuttle in 2010 nonetheless, however, and reportedly is assessing other methods for launching ISS segments.

"Expedition" crews have occupied ISS on a 4-6 month rotating basis since November 2000. Originally the crews had three members (two Russians and one American, or two Americans and one Russian), with an expectation that crew size would grow to six or seven once assembly was completed. Crew size is temporarily reduced to two (one American, one Russian) while the U.S. shuttle is grounded in order to reduce resupply requirements. The number of astronauts who can live on the space station is limited in part by how many can be returned to Earth in an emergency by lifeboats docked to the station. Only Russian Soyuz spacecraft are available as lifeboats. Each Soyuz can hold three people, limiting crew size to three if only one Soyuz is attached. NASA planned to build a U.S. Crew Return Vehicle (CRV) to provide lifeboat capabilities for at least four more crew. The Bush Administration canceled those plans due to cost growth in the ISS program, then began a different program (the Orbital Space Plane) that also was cancelled. In September 2005, NASA announced that the new Crew Exploration Vehicle (CEV) it is building to implement the President's Vision for Space Exploration (the "Moon/Mars" program) will be designed to take crews to and from the ISS, and to serve as a lifeboat. NASA currently hopes to have it ready by 2012.

Each Soyuz must be replaced every six months. The replacement missions are called "taxi" flights since the crews bring a new Soyuz up to ISS and bring the old one back to Earth. Therefore, under normal conditions, the long duration Expedition crews are regularly visited by taxi crews, and by the space shuttle bringing up additional ISS segments or exchanging Expedition crews. When the shuttle is unavailable, Expedition crews are taken back and forth on the "taxi" flights.

September 1993-January 2001: The Clinton Administration

Cost Growth

From FY1994-FY2001, the cost estimate for building ISS grew from \$17.4 billion to \$24.1-26.4 billion. The \$17.4 billion estimate (called its "development cost," "construction cost," or "R and D cost") covered FY1994 through completion of assembly, then scheduled for June 2002. It did not include launch costs, operational costs after completion of assembly, civil service costs, or other costs. NASA estimated the program's life-cycle cost (all costs, including funding spent prior to 1993) from FY1985 through FY2012 at \$72.3 billion. In 1998, GAO estimated the life-cycle cost at \$95.6 billion (GAO/NSIAD-98-147). More recent, comparable, life-cycle estimates are not available from NASA or GAO.

Cost growth first emerged publicly in March 1996 when then-NASA Administrator Daniel Goldin gave the space station program manager control of money allocated for (and previously overseen by) the science offices at NASA for research intended to be conducted aboard the space station. Congress gave NASA approval to transfer \$177 million from those science accounts to space station construction in the FY1997 VA-HUD-IA appropriations act (P.L. 104-204). A similar transfer was approved for FY1996 (\$50 million). NASA changed its accounting methods so future transfers would not require congressional action, and transferred \$235 million from space station science into construction in FY1998. ("Space station science" funding is for scientific activities aboard the space station. It is separate from NASA's other "space science" funding, such as Mars exploration.)

One factor in the cost growth was schedule slippage related to Russia's Zvezda module. As insurance against further Zvezda delays, or a launch or docking failure, NASA decided to build an "Interim Control Module" (ICM). To cover cost growth associated with the schedule delay and ICM, NASA requested permission to move $200 million in FY1997 from the space shuttle and payload utilization and operations accounts to the space station program, and to transfer $100 million in FY1998 from unidentified NASA programs to the space station program. The appropriations committees approved transferring the $200 million in FY1997, but not the FY1998 funding.

In March 1998, NASA announced that the estimate for building the space station had grown from $17.4 billion to $21.3 billion. In April 1998, an independent task force concluded that the space station's cost through assembly complete could be $24.7 billion. Mr. Goldin initially refused to endorse the $24.7 billion estimate, but by 2000, NASA's own estimate had grown to $24.1-$26.4 billion.

Cost Caps

The $2.1 billion per year figure the Clinton White House and Congress agreed to spend on the space station, and NASA's $17.4 billion estimate to build the station, became known as "caps," although they were not set in law. Both were exceeded in 1997-1998. As costs continued to rise, Congress voted to legislate caps on certain parts of the ISS program in the FY2000-2002 NASA authorization act (P.L. 106-391). The caps are $25 billion for development, plus $17.7 billion for associated shuttle launches. The act also authorizes an additional $5 billion for development and $3.5 billion for associated shuttle launches in case of specified contingencies. The caps do not apply to operations, research, or crew return activities after the space station is "substantially" complete, defined as when development costs consume 5% or less of the annual space station budget. In its FY2006 budget justification (p. EC 2-4), NASA alerted Congress that it might exceed the $25 billion cap for ISS development during FY2005, attributing the increased costs to delays resulting from the *Columbia* tragedy. The House version of the pending NASA authorization bill, H.R. 3070, would repeal the cap. The Senate version, S. 1281, would require NASA to submit a report on how certain factors affected ISS development costs, and to identify statutory changes needed to address those impacts. NASA did exceed the cap in FY2005, having spent $25.7 billion in funds that are counted against the cap.

2001-Present: The George W. Bush Administration

Cost Growth

As President Bush took office, NASA revealed substantial additional cost growth. In 2000, NASA's estimate of the remaining cost to build ISS was $8 billion (FY2002 to FY2006). In January 2001, however, it announced that an additional $4.02 billion was needed. That figure grew to $4.8 billion by June, and the IMCE task force (discussed below) said another $366 million in growth was discovered between August and October. Those increases would have raised the cost to over $30 billion, 72% above the 1993 estimate, and $5 billion above the legislated cap. NASA explained that program managers had underestimated the complexity of building and operating the station. The Bush Administration signaled it supported the legislated cap, would not provide additional funds, and NASA would have to find what it needed from within its Human Space Flight account.

"Core Complete" Configuration

In February 2001, the Bush Administration announced it would cancel or defer some ISS hardware to stay within the cap and control space station costs. It canceled the Propulsion Module, and "indefinitely deferred" the Habitation Module, Node 3, and the Crew Return Vehicle (CRV). The decision truncated construction of the space station at a stage the Administration called "core complete." The Administration said that "enhancements" to the station might be possible if NASA demonstrated improved cost estimating and program management. In 2001, the space station program office at Johnson Space Center (JSC) estimated that it would cost $8.3 billion from FY2002-2006 to build the core complete configuration, described at that time as all the U.S. hardware planned for launch through "Node 2," plus the launch of laboratories being built by Europe and Japan. NASA subsequently began distinguishing between "U.S. Core Complete" (the launches through Node 2, which, prior to the *Columbia* tragedy, was scheduled for February 2004) and "International Partner (IP) Core Complete" which included the addition of European and Japanese laboratory modules (then anticipated in 2008).

The non-U.S. partners, and U.S. scientists who planned to conduct research on ISS, expressed deep concern with the core complete configuration (see CRS Report RL31216). A major issue was that NASA reduced its space station research budget by 37.5% over the FY2002-2006 period, necessitating a reassessment of U.S. research priorities on ISS. A July 2002 report of the "Research Maximization and Prioritization" (ReMaP) task force, and a

September 2002 National Research Council report, made recommendations on research priorities. Both were superseded by President Bush's January 2004 "Vision for Space Exploration," which directs that U.S. research on ISS be restricted only to that which supports the Vision. A new research plan has not been released by NASA. At the time the core complete configuration was announced, another major concern was the decision to indefinitely defer the CRV, which subsequently was canceled. That would have limited the space station to three permanent crew members, not seven as planned, reducing the number of researchers on board to conduct the research program. As discussed elsewhere in this report, NASA reinstated plans for a CRV capability in 2005.

Reviews of NASA's Cost Estimates and Adding Funds for ISS

NASA created the ISS Management and Cost Evaluation (IMCE) Task Force in July 2001 to review the space station program office's $8.3 billion cost estimate for finishing the core complete configuration. Chaired by former Lockheed Martin executive Tom Young, IMCE determined that the cost estimate was not credible, and NASA should make significant management and cost estimating changes (see CRS Report RL31216). NASA Headquarters directed the space station program office to reassess its estimate, and had two independent groups conduct their own estimates. A July 2002 GAO report (GAO-02-735) concluded that NASA's focus on managing annual budgets resulted in NASA's failure to heed indicators of future program cost growth.

In November 2002, the Bush Administration submitted an amended FY2003 budget request that shifted $706 million into the ISS program for FY2004-2007: $660 million to boost program reserves, and $46 million in FY2004 for "long-lead" items to preserve the option of increasing crew size beyond three. (Congress cut $200 million from ISS in FY2004, however.) The latter included a proposal to build an Orbital Space Plane (OSP) to takes crews back and forth to ISS as a complement to the space shuttle.

At a December 2002 "Heads of Agency" meeting, the ISS partners agreed on a process for selecting a final ISS configuration by December 2003. The 2003 space shuttle *Columbia* tragedy delayed the process, and President Bush's January 2004 announcement of the Vision for Space Exploration, changed NASA's own plans, including cancellation of the Orbital Space Plane, and termination of the space shuttle program after ISS construction is completed. At a January 2005 Heads of Agency meeting, the partners endorsed a final configuration of ISS, but NASA subsequently announced changes to it. The agency now plans to conduct only 18 (instead of 28) launches to the ISS, all before the end of 2010, and has dropped plans to launch the centrifuge and its accommodation module, and Russia's Science Power Platform. The agency plans to meet with the other ISS partners to discuss these changes in late 2005.

The changes to the ISS are largely due to the new direction NASA is taking in response to the Vision for Space Exploration. Inter alia, the Vision calls for development of a Crew Exploration Vehicle (CEV) to take astronauts to and from the Moon. It also can take them to and from the ISS, and NASA Administrator Griffin stated at a September 19, 2005 press conference that the CEV would be used to take crews to and from the ISS, and to serve as a lifeboat for them. If the CEV is built as announced, it would fulfill the U.S. commitment to build a crew return capability, and allow the ISS crew size to increase to its originally planned complement of seven. President Bush directed that the CEV be ready by 2014; Dr. Griffin hopes to accelerate the schedule to 2012.

CONGRESSIONAL ACTION

FY2005

The FY2005 request for the ISS program was $2.412 billion: $1.863 billion for construction and operations, including $140 million in a new "ISS Crew/Cargo Services" line to pay for alternatives to the shuttle for taking crew and cargo to and from ISS; and $549 million for research. Congress did not specify a funding level for the ISS in the final version of the FY2005 VA-HUD-IA appropriations act, which was incorporated in the FY2005 Consolidated Appropriations Act (H.R. 4818, P.L. 108-447). Instead, it gave NASA "unrestrained transfer authority" to shift money between budget accounts. In a May 10, 2005 update to its FY2005 operating plan, NASA indicated that it is shifting $160 million from the space station into the space shuttle program for costs associated with returning the shuttle to flight status. The operating plan shows $1.676 billion for ISS construction and operations, including $98 million for ISS Crew/Cargo Services. However, management of the ISS Crew/Cargo Services activity, and the $98 million, have been moved to the Exploration Systems Mission Directorate. The operating plan retains the $98 million in the ISS subaccount, however. The FY2005 total shown in the table in this issue brief — $2.058 billion — is the sum of $1.676 billion as shown in NASA's May 2005 operating plan, plus $382 million allocated for space station research as shown in a FY2006 NASA budget chart.

Table 1. U.S. Space Station Funding (in $ Millions)

Fiscal Year	Request	Appropriated
1985	150	150
1986	230	205
1987	410	410
1988	767	425
1989	967	900
1990	2,050	1,750
1991	2,430	1,900
1992	2,029	2,029
1993	2,250	2,100
1994	2,106	2,106
1995	2,113	2,113
1996	2,115	2,144
1997	2,149	2,149
1998	2,121	2,441[A]
1999	2,270	2,270
2000	2,483	2,323
2001	2,115	2,115
2002	2,114	2,093
2003	1,839	1,810 **B**
2004 **C**	2,285	2,085
2005	2,412	2,058 **D**
2006	1,995	

The numbers here reflect NASA's figures for "the space station program." Over the years, what is included in that definition has changed. The appropriated amount may differ from actual spending.

[a] NASA's FY1999 budget documents showed $2.501 billion in the expectation Congress would approve additional transfer requests, but it did not.

[b] Adjusted for 0.65% rescission.

[c] Reflects shift to full cost accounting.

[d] Congress did not specify an appropriations level. This figure is from NASA's May 10, 2005 operating plan, plus the amount to be spent on ISS research.

FY2006

For FY2006, NASA originally requested $2.180 billion for the ISS program: $1.857 billion for construction and operations (including $160 million for ISS Crew/Cargo Services), and $324 million for ISS research. In a May 10, 2005 FY2005 operating plan update, NASA announced that it was moving the ISS Crew/Cargo Services activity to the Exploration Systems Mission Directorate (ESMD). A July 15 amended budget request also moved the FY2006 funding for that activity (which it said was $168 million) into ESMD and commensurately reduced the ISS construction and operations request. NASA later also reduced the amount it plans to spend on ISS research to $306 million. The number used in the accompanying table ($1,995 million) reflects these changes.

The FY2006 Science, State, Justice and Commerce (SSJC) appropriations bill (H.R. 2862), includes NASA. The House cut $10 million from ISS construction and operations, $10 million from ISS Crew/Cargo Services, and $25 million from the account that funds research on ISS, though it did not specify that the cuts come from the ISS portion of that account. The Senate cut all $160 million from ISS Crew/Cargo Services. The final version of the bill cut $80 million, including $60 million from ISS Crew/Cargo Services (the conference report does not reflect the July budget amendment.) S. 1281, a FY2006-2010 NASA authorization bill, adds $100 million in FY2006 and makes other changes to enhance ISS research. H.R. 3070 (H.Rept. 109-173) is a FY2006-2007 NASA authorization bill, that does not specify an amount for ISS.

INTERNATIONAL PARTNERS

The Original Partners: Europe, Canada, and Japan

Canada, Japan, and most of the 16 members of the European Space Agency (ESA) have been participating in the space station program since it began. Formal agreements were signed in 1988, but had to be revised following Russia's entry into the program, and two more European countries joined in the interim. The revised agreements were signed on January 29, 1998, among the partners in the ISS program: United States, Russia, Japan, Canada, and 11 European countries — Belgium, Denmark, France, Germany, Italy, the Netherlands, Norway, Spain, Sweden, Switzerland, and the United

Kingdom. Representatives of the various governments signed the government-to-government level Intergovernmental Agreement (IGA) that governs the program. (The United Kingdom signed the IGA, but is not financially participating in the program so the number of European countries participating in the program is variously listed as 10 or 11.) NASA also signed Memoranda of Understanding for implementing the program with its counterpart agencies: the European Space Agency (ESA), the Canadian Space Agency (CSA), the Russian space agency (then Rosaviakosmos, now Roskosmos), and the Japanese Science and Technology Agency. The IGA is a treaty in all the countries except the United States (where it is an Executive Agreement).

Canada is contributing the Mobile Servicing System (MSS) for assembling and maintaining the space station. In February 1994, the new prime minister of Canada had decided to terminate Canada's role in the program, but later agreed to reformulate Canada's participation instead. The first part of the MSS (the "arm" or Canadarm 2) was launched in April 2001; another part, the Special Purpose Dextrous Manipulator, is awaiting launch.

ESA is building a laboratory module called Columbus, and an Automated Transfer Vehicle (ATV) to take cargo to ISS. The ATV will be launched on Europe's Ariane launch vehicle. The first ATV launch is expected in 2006. The major contributors to Columbus are Germany, France, and Italy. Budgetary difficulties over the years led ESA to cancel other hardware it was planning. ESA also is building a cupola (a windowed dome) and two of the three "nodes" (Node 2 and Node 3) for NASA in exchange for NASA launches of Europe's module and other services. Node 2, Columbus, and the cupola are awaiting launch. NASA had canceled plans for Node 3, but now has revived them. NASA also has a bilateral agreement with Italy under which Italy built three "mini-pressurized logistics modules" (MPLMs). Already in use, they are launched via the shuttle, attached to ISS while cargo is transferred to the station, filled with refuse or other unwanted material, placed back into the shuttle's cargo bay, and returned to Earth.

Japan is building the Japanese Experiment Module, named Kibo (Hope). One part is pressurized and the other is not (called the "back porch," it will be exposed to space for experiments requiring those conditions). The pressurized section is awaiting launch; the unpressurized section is in development. Japan also is building a centrifuge and a Centrifuge Accommodation Module ("CAM") for NASA in exchange for shuttle flights to launch Kibo, but NASA decided in September 2005 that it no longer needs the centrifuge. NASA plans to meet with Japan and the other partners in late 2005 to discuss this and other matters.

CSA reported in February 2004 that Canada's total ISS spending is expected to be $1.4 billion (Canadian), of which $1.3 billion (Canadian) was spent by that time. ESA reported in March 2004 that its estimated funding for ISS is 5.1 billion Euros, of which 4.1 billion Euros were spent as of the end of 2003. (In March 2004, 1 Canadian dollar = 0.75 U.S. dollar and 1 Euro = 1.2 U.S. dollars.) In February 2004, the Japanese space agency reported that Japan expects to spend $4.8 billion on ISS, of which $4 billion was spent by the end of March 2003. (A bilateral agreement was signed with Brazil in October 1997 for Brazil to provide payload and logistics hardware. Brazil is restructuring its agreement in light of financial constraints, however. The level of its funding contribution is unclear.)

Russia

Issues associated with Russia's participation in ISS are discussed elsewhere. This section explains Russian space station activities from 1971 to the present. The Soviet Union launched the world's first space station, Salyut 1, in 1971 followed by five more *Salyuts* and then *Mir*. At least two other *Salyuts* failed before they could be occupied. The Soviets accumulated a great deal of data from the many missions flown to these stations on human adaptation to weightlessness. The data were often shared with NASA. They also performed microgravity materials processing research, and astronomical and Earth remote sensing observations. Importantly, they gained considerable experience in operating space stations. Russia's most recent space station, *Mir*, was a modular space station built and operated between 1986 and 2001. Crews were ferried back and forth to *Mir* using Soyuz spacecraft.

Crews occupied *Mir* from 1986-2000. For almost ten of those years (1989-1999), *Mir* was continuously occupied by crews on a rotating basis. Although occasionally crews stayed for very long periods of time to study human reaction to long duration spaceflight, typically they remained for 5-6 months and then were replaced by a new crew. From 1995-1998, seven Americans participated in long duration (up to six month) missions aboard *Mir*, and nine space shuttle missions docked with the space station. Individuals from Japan, Britain, Austria, Germany, France, and the Slovak Republic also paid for visits to *Mir*. Russia deorbited *Mir* into the Pacific Ocean on March 23, 2001.

Risks and Benefits of Russian Participation

For many years, controversy over the ISS program focused on Russia's participation in the program. Among the issues were the extent to which successful completion of ISS is dependent on Russia, Russia's financial ability to meet its commitments, and whether the United States should provide funding to Russia if it proliferates missile technology to certain countries. While there is no exchange of funds among the other ISS partners, the United States (and other partners) have provided funding to Russia. By 1998, the United States had paid approximately $800 million to Russia for space station cooperation.

Following the Clinton Administration's decision to bring Russia into the program, Congress stated that Russian participation "should enhance and not enable" the space station (H.Rept. 103-273, to accompany H.R. 2491, the FY1994 VA-HUD-IA appropriations act, P.L. 103-124). The current design, however, can only be viewed as being "enabled" by Russian participation. It is dependent on Russian Progress vehicles for reboost (to keep the station from reentering Earth's atmosphere), on Russian Soyuz spacecraft for emergency crew return, and on Russia's Zvezda module for crew quarters (which allows ISS to be permanently occupied). When the shuttle is unavailable, U.S. access to ISS is completely dependent on Russia, which ferries crews back and forth on the Soyuz spacecraft and takes cargo to ISS on Progress spacecraft. President Bush's exploration initiative would increase U.S. dependence on Russia vis a vis the space station (see Key Issues for Congress below).

Russia's financial ability to meet its commitments is an ongoing issue. The launch of Zvezda, the first module Russia had to pay for itself, was more than two years late. (Zarya was built by Russia, but NASA paid for it.) Since Zvezda's launch in 2000, Russia has met its commitments to launch Soyuz and Progress spacecraft, but is reassessing what other modules and hardware it will build at its own expense. Russian space agency officials have repeatedly expressed concern about whether they can provide the needed number of Soyuz and Progress spacecraft because of budget constraints.

Clinton Administration and NASA officials asserted repeatedly that Russian participation in the space station program would accelerate the schedule by two years and reduce U.S. costs by $4 billion. That was later modified to one year and $2 billion, and an April 1, 1994 letter to Congress from NASA said 15 months and $1.5 billion. NASA officials continued to use the $2 billion figure thereafter, however. GAO concluded (GAO/NSIAD 94-248) that Russian participation would cost NASA $1.8 billion, essentially negating the $2 billion in expected savings. In 1998, a NASA official conceded that having Russia as a partner added $1 billion to the cost. Other benefits cited by the Clinton Administration were providing U.S. financial assistance to Russia as it moves to a market economy, keeping Russian aerospace workers employed in non-threatening activities, and the emotional impact, historic symbolism, and potential long term significance of the two former Cold War adversaries working together in space.

One benefit is that the space station can be serviced with Russian as well as American spacecraft, providing redundancy if either side must ground its fleet due to an accident, for example. This is an important advantage while the U.S. space shuttle is grounded. Russia is providing both crew and cargo flights to the space station, enabling it to continue operation without the shuttle.

ISS and U.S. Nonproliferation Objectives, Including the Iran Nonproliferation Act (INA)

The overall relationship between the United States and Russia is another factor in the ISS equation, including Russian adherence to U.S. nonproliferation objectives. Getting Russia to adhere to the Missile Technology Control Regime (MTCR), designed to stem proliferation of ballistic missile technology, appears to have been a primary motivation behind the Clinton Administration's decision to add Russia as a partner. The United States wanted Russia to restructure a contract with India that would have given India advanced rocket engines and associated technology and know-how. The United States did not object to giving India the engines, but to the technology and know-how. Russia claimed that restructuring the contract would cost $400 million. The 1993 agreement to bring Russia into the space station program included the United States paying Russia $400 million for space station cooperation. At the same time, Russia agreed to adhere to the MTCR. The question is what the United States will do if Russia violates the MTCR. Some Members of Congress believe Russia already has done so. The Clinton Administration sanctioned 10 Russian entities for providing technology to Iran. Neither the Russian space agency nor any major Russian ISS contractors or subcontractors were among those sanctioned.

On March 14, 2000, President Clinton signed into law the Iran Nonproliferation Act (INA), P.L. 106-178. The law, *inter alia*, prohibits NASA from making "extraordinary payments" related to ISS after January 1, 1999, in cash or in kind, to Russia unless Russia takes the necessary steps to prevent the transfer of weapons of mass destruction and missile systems to Iran, and the President makes a determination that neither the Russian space agency nor any entity reporting to it has made such transfers for at least one year prior to such determination. Exceptions are made for payments needed to prevent imminent loss of life by or grievous injury to individuals aboard ISS (the "crew safety" exception); for payments to construct, test, prepare, deliver, launch, or maintain Zvezda as long as the funds do not go to an entity that may have proliferated to Iran and the United States receives goods or services of commensurate value; and hardware needed to dock the U.S. Interim Control Module (ICM, discussed earlier). Certain notifications are required if the exceptions are utilized. NASA was seeking permission to spend $35 million on Russia goods and services, of which $14 million was for the ICM docking hardware. President Clinton provided Congress with a required notification with regard to that $14 million on June 29, 2000. Ultimately, only $11 million was needed for the ICM hardware, leaving $24 million that NASA wanted to spend. No determination as required by the act was forthcoming from the President. NASA considered using the crew safety exception, but at a House International Relations Committee hearing on October 12, 2000, some committee Members sharply criticized NASA's legal interpretation of that exception, particularly NASA's broad interpretation of the word "imminent."

Thus, the INA has important ramifications for whether NASA can keep its astronauts on ISS for long duration missions after April 2006, or at all after 2010 if the shuttle is terminated as planned (see Key Issues for Congress, below). The Bush Administration submitted a proposed amendment to Congress on July 12, 2005, to modify the INA. The amendment would have prohibited payments only for goods or services that Russia had previously agreed to provide at no cost. On September 21, 2005, the Senate passed S. 1713 (Lugar) which would have allowed NASA to pay Russia, but only through January 1, 2012. The House passed an amended version on October 26 that, inter alia, clarifies that the goods or services must be delivered, as well as paid for, by January 1, 2012. (The House version makes other changes as well that are outside the scope of this report to discuss.) The Senate agreed to the House-passed version on November 8. The bill is awaiting signature by the President. See CRS Report RS22270 for more information.

KEY ISSUES FOR CONGRESS

Maintaining ISS Operations while the Shuttle Is Grounded

The grounding of the space shuttle system following the *Columbia* accident suspended assembly of ISS, temporarily reduced the size of Expedition crews from three to two, and complicated efforts to keep the crews supplied with consumables, scientific experiments, and spare parts for equipment that needs repair. NASA launched the shuttle on its first Return to Flight mission, STS-114, on July 26, 2005. STS-114 landed successfully on August 9, after spending much of its mission docked with ISS. Another shuttle launch was scheduled for September, but a problem occurred during STS-114's launch and NASA has again grounded the fleet. The next shuttle mission is not expected until at least May 2006.

In the absence of the shuttle, ISS crews are taken to and from ISS using Russian Soyuz spacecraft on the same six-month schedule already planned, and Russian Progress spacecraft resupply the crew. Russia is obligated to provide crew return for three people throughout the lifetime of ISS. Currently, they accomplish that with two Soyuzes per year (each lasts only six months once docked to ISS). Russia also is obligated to provide a certain number of Progress spacecraft, but has cautioned that funding for Soyuz and Progress is not assured.

The Russians operated seven of their own space stations (see above) using only Soyuz and Progress spacecraft, so it is possible to keep ISS operating without the shuttle as long as Russia is willing to provide them. However, operation of ISS was premised on the availability of the cargo-carrying capacity of the space shuttle. Keeping ISS operating without the shuttle is challenging. For example, the Expedition 10 crew was required to reduce its food intake because of shortages aboard the station in late 2004. Stocks were resupplied by a Progress that reached ISS late that December, but U.S. and Russian space station personnel made clear that if the Progress had failed to dock, the crew would have had to return home prematurely because of the food situation. ISS crews also need to repair faulty equipment, but replacement parts may not fit aboard Progress or Soyuz. NASA's decision to again ground the shuttle fleet may lead to questions about whether to keep a crew aboard ISS. In addition to questions about keeping the crews well supplied, with a two-person crew, less time may be available for scientific research, and without the shuttle, fewer experiments can be taken to ISS. If little scientific research can be accomplished, some may question the wisdom of asking astronauts and cosmonauts to accept the risks inherent in human spaceflight simply to maintain ISS systems. Conversely, how long ISS could continue to function with no one aboard is unknown. Progress spacecraft could dock with ISS automatically to reboost it and keep it at the proper altitude, but a major system malfunction that could not be remedied by ground-based controllers could imperil the station.

Ensuring U.S. Astronaut Participation in Long-Duration Missions

As noted, ISS crew size is limited in part by the number of occupants that can be accommodated in a lifeboat, or "crew return vehicle" (CRV), in an emergency. Without its own CRV until at least 2012, NASA faced two deadlines in terms of assuring access to ISS for long duration missions by U.S. astronauts. First was spring 2006, when Russia will have fulfilled its commitment to provide lifeboat services for U.S. astronauts. Second was 2010, when the shuttle is expected to be retired in accordance with President Bush's Vision for Space Exploration. The 2010 deadline is discussed separately (see next issue).

Regarding the spring 2006 deadline, the international agreements that govern the ISS program obligate Russia to provide crew return services for three crew members throughout the lifetime of the ISS, and the United States to provide such services for at least four people beginning when assembly of ISS is completed. Prior to the *Columbia* tragedy, that milestone was expected to occur in 2006. By that time, a U.S. CRV, accommodating at least four people, was expected to be available. It would not only provide lifeboat capabilities for U.S. crew members, but allow the ISS crew size to grow to seven (three could return on a Soyuz, and four on the CRV), including representatives from the other partners. The Bush Administration terminated the CRV, and its successor, the Orbital Space Plane. NASA now has resumed planning to build a spacecraft that could serve this function (the Crew Exploration Vehicle), but it is not expected to be ready until at least 2012. Therefore, only Russian Soyuz spacecraft are currently available for lifeboat services.

Under a 1996 "balance agreement" between NASA and the Russian space agency, Russia is obligated to use 11 Soyuz spacecraft to provide crew return for U.S. crews. The last of the 11 was launched on October 1, 2005, and is expected to return to Earth in April 2006. After that, Russia no longer must allocate any of the seats on its Soyuzes for U.S. astronauts. It can sell those flight opportunities to whomever it wishes. Russia is interested in selling flight opportunities to help finance its space program, so might choose to limit those opportunities to paying customers. As noted, the INA prevents NASA from paying Russia for ISS-related activities unless Russia does not proliferate certain technologies to Iran. Congress has passed legislation (S. 1713) that allows NASA to purchase ISS-related goods and services through January 1, 2012. The bill is awaiting signature by the President.

Impact of President Bush's Vision for Space Exploration, Including a Potential Gap in U.S. Human Access to Space

President Bush's January 2004 Vision for Space Exploration directs NASA to focus its activities on returning humans to the Moon by 2020, and someday sending them to Mars and "world beyond." The Vision affects the ISS program in several ways. First, the President directed that the shuttle be retired in 2010, and NASA officials indicated that they would complete their use of the ISS in 2016. By terminating the shuttle and NASA utilization of the ISS, that funding (approximately $6 billion per year) could be reallocated to achieving other aspects of the Vision. The President also directed that the U.S. research on ISS be restricted to research needed to support the Vision, instead of the broadly-based research program that was planned.

The extent to which ISS can be utilized without the space shuttle is not clear. Soyuz spacecraft can take crews back and forth, but the shuttle's cargo capacity — both for taking cargo to ISS, and back to Earth (e.g. the results of scientific experiments, or hardware that needs repair) — could be expensive to replicate. No other partner has a spacecraft able to bring material back to Earth today. Europe reportedly is considering adding a return capsule to its Automated Transfer Vehicle (a robotic cargo spacecraft that is expected to make its first flight in 2007). NASA is exploring the possibility of U.S. commercial companies providing launch and/or return services through the ISS Crew/Cargo activity.

Another issue was how U.S. crews could access ISS once the shuttle is retired in 2010. As already discussed, NASA was facing an April 2006 deadline regarding Russian support for U.S. astronauts because of the Iran Nonproliferation Act (INA). That issue appears to be resolved with congressional passage of S. 1713, though that allows NASA to purchase Russian goods and services only through January 1, 2012. If NASA's new

CEV is not available by that time, additional modifications to the INA may be required. Once the shuttle is retired in 2010, NASA will not be able to launch its own astronauts into space at all until the CEV is ready.

LEGISLATION

H.R. 2862 (Wolf). FY2006 Science, State, Justice, Commerce Appropriations Act (includes NASA). Reported from House Appropriations Committee June 10, 2005 (H.Rept. 109-118); passed House June 16. Reported from Senate Appropriations Committee June 23 (S.Rept. 109-88); passed Senate September 15, 2005. Conference report (H.Rept. 109-272) filed November 7, passed House November 9, passed Senate November 16.

H.R. 3070 (Calvert)/S. 1281 (Hutchison). NASA authorization bill. H.R. 3070 authorizes NASA funding for FY2006-2007; reported by the House Science Committee July 18 (H.Rept. 109-173), passed House, amended, July 22, 2005. S. 1281 authorizes NASA funding for FY2006-2010; reported by Senate Commerce Committee July 26 (S.Rept. 109-108), passed Senate, amended, September 28, 2005.

S. 1713 (Lugar). Amends the Iran Nonproliferation Act to allow NASA to purchase ISS-related services from Russia through January 1, 2012. Discharged from Senate Foreign Relations Committee and passed Senate September 21. Passed House, amended, October 26, 2005. Senate agreed with House-amended version, November 8. Presented to the President November 17.

CHAPTER SOURCES

The following chapters have been previously published:

Chapter 1 – This is an edited, excerpted and augmented edition of a United States Congressional Budget Office, a CBO Study, dated October 2006.

Chapter 2 – This is an edited, excerpted and augmented edition of a United States Government Accountability Office (GAO) Report to the Subcommittee on Defense, Committee on Appropriations, House of Representatives. Publication GAO-08-1039, dated September 2008.

Chapter 3 – This is an edited, excerpted and augmented edition of Congressional Research Service Report IB93017, dated November 17, 2005.

Chapter 4 - This is an edited, excerpted and augmented edition of Congressional Research Service Report IB93062, dated March 20, 2006.

INDEX

A

B

C

D

E

F

G

H

I

J

K

L

M

Q

R

S

T

U

V

W

Y